P9-CDK-518

ON BEING A BEAR

On Being a Bear

FACE TO FACE WITH OUR WILD SIBLING

RÉMY MARION

Translated by David Warriner

Foreword by Lambert Wilson

GREYSTONE BOOKS
Vancouver/Berkeley

First published in English by Greystone Books in 2021
Originally published in French as *L'ours: L'autre de l'homme,*
copyright © 2018 by Actes Sud, Arles, France
English translation copyright © 2021 by David Warriner

21 22 23 24 25 5 4 3 2 1

All rights reserved. No part of this book may be reproduced, stored in a retrieval system or transmitted, in any form or by any means, without the prior written consent of the publisher or a license from The Canadian Copyright Licensing Agency (Access Copyright). For a copyright license, visit accesscopyright.ca or call toll free to 1-800-893-5777.

Greystone Books Ltd.
greystonebooks.com

Cataloguing data available from Library and Archives Canada
ISBN 978-1-77164-698-7 (cloth)
ISBN 978-1-77164-699-4 (epub)

Copy editing by Paula Ayer
Proofreading by Alison Strobel
Indexing by Stephen Ullstrom
Cover and text design by Fiona Siu
Maps by Pascal Orcier
Author photo by Thierry Borredon

Printed and bound in Canada on ancient-forest-friendly paper by Friesens

Greystone Books gratefully acknowledges the Musqueam, Squamish, and Tsleil-Waututh peoples on whose land our office is located.

Greystone Books thanks the Canada Council for the Arts, the British Columbia Arts Council, the Province of British Columbia through the Book Publishing Tax Credit, and the Government of Canada for supporting our publishing activities.

Canada

Canada Council for the Arts
Conseil des arts du Canada

For Vadim and Solal, my grandsons

Contents

Foreword

BEAR,

I owe you an apology.

It turns out I knew nothing about you. Not before I read this captivating account of your life by Rémy Marion. Until then, I had fallen for all the clichés. You were a gentle, hungry giant, an endearing, clumsy clown. You and your kind had none of the elegance or mystery of the feline species, and none of their superior indifference. No—you were a dull-witted, boorish glutton, but you were also a dangerous, voracious beast.

I may have lent you my voice in the movie theater when I was cast as Baloo in the French version of *The Jungle Book* film in 2016, and when I played Ernest the bear, companion to Celestine the mouse, in the animated film *Ernest & Celestine*, but I couldn't see that I had reduced you to a theme-park caricature.

This book has opened my eyes. Rémy Marion has followed you, spied on you, analyzed you, photographed you, and filmed you for years. He knows you inside out, and he never tires of observing you, watching out for you, waiting for you in every corner of the planet where you have survived.

You know, Bear, you and I are not so different. Sometimes I'm afraid of people too. I want to run away from them just like you do. Once you lived alongside these people and they looked up to you like a god, but now they have hunted you, enslaved you, and invaded your territory. I don't blame you for keeping your distance. They can be dangerous. Not all of them, of course, but a lot of them are out to get you. They want your hide. Bears are good for nothing anymore, they say. They are a nuisance, so we have to get rid of them. And get rid of you they will, the same way they do with every kind of creature that stands in their way. I hate to tell you, Bear, but your days are numbered.

On behalf of the people, on behalf of all humankind, I would like to offer you the sincerest and most meaningful of apologies. Sorry for driving you away to places where you struggle to find food. Sorry for turning you into countless bedside rugs. Sorry for locking you up in cages to drain you of your bile, the way people still do in some parts of the world. Sorry for melting your pack ice. Sorry for pushing you to the edge of starvation, leaving you to fill your belly in our landfills. Sorry for all those hunts in the Pyrenees, where the shepherds' hatred for you dates back to the time of their ancestors (were you not, however, in those mountain pastures well before their flocks?). Sorry for dressing you up like a circus animal, putting a ring through your nose, and making you dance with monkeys.

Still—not that you know it—you've become something of a symbol, an emblem of lost harmony with the humans in whose midst you've managed to survive for thousands of years. Yet soon you'll be swept away by the black tide of their excesses and caught in the trap of their greed and foolishness.

Don't let it go to your head, Bear, you're no more lovable than any of the other species in this world. You're simply a part of the wonderful diversity of our planet that is so terribly fragile and likely headed for extinction. Unless...

Unless people choose to follow the path Rémy Marion has forged.

He loves you. He holds you dear to his heart—from a certain distance, though sometimes he gets a little too close for your liking! His admiration for you is limitless, as is his fascination with you. But most of all, above all else, he respects you. I hope this fantastic book of his will open people's eyes and help them to see who you truly are, so they can finally give you the love and respect you deserve.

LAMBERT WILSON

A magnificent female brown bear, fishing on Russia's Kamchatka Peninsula

Introduction

Bears are made of the same dust as we,
and they breathe the same winds
and drink of the same waters.

John Muir[1]

IT MAY NOT surprise you that my fascination with wild life and my drive to travel the world and explore the great wide open are rooted in my childhood.

As a young boy, I used to watch my father head out fishing every night. In the winter, he would don his heavy peacoat, woolen scarf, and navy blue cloth hat. After listening religiously to the shipping forecast on the radio, he would set off on an adventure to some unknown destination. Or that's the way it seemed to me. Utsire, Dogger, FitzRoy, Biscay, Dover, and Wight... the names of those places are still as fresh in my mind as the images of the fog, the fish, and the waves I conjured up. He never went that far, nowhere near those distant stormy seas, but the Baie de Seine, where the river that flows through Paris meets the English Channel, is far more treacherous than any open waters.

In the morning, after my father had returned, I'd find live shrimp in a bowl or hermit crabs in a bucket that he had caught and brought back for me play with. These poor creatures, plucked so swiftly out of their element, told me tales of the tides, the currents, and the watery depths.

My childhood was a long voyage of discovery, spent searching for marine creatures and wandering the beaches of Honfleur alone. I didn't know it then, but these activities were laying the early foundations for my expeditions to the Far North. The ever-changing light in the Baie de Seine was a source of inspiration for the Impressionist painters, and it inspired me too. Having feasted my eyes on paintings by Marie Laurencin, Eugène Boudin, Claude Monet, Henri de Saint-Delis, and André Hambourg from a very young age, I developed a keen appetite for temperamental skies, clouds, and waves.

Years went by before I saw my first bear, before I roamed the pack ice and boreal forest, but clearly I had been bitten by the bug back then, in my youth. As the Arctic explorer Jean-Baptiste Charcot wrote:

> These polar regions hold such a strange attraction, so powerful, so persistent, that upon one's return one tends to forget the physical and moral fatigue of it all, and think only of going back there. Why, I wonder? [...] Any man fortunate to enter such a place shall feel his spirits lifted.[2]

Since the late 1980s, I've traveled far and wide several times a year to observe, study, photograph, and film bears with brown, white, and black fur. These encounters with nature have often been exceptional, but they've also been a very human kind of experience, as I've rubbed shoulders with guides, hunters, scientists, and geographers, all of whom feel as passionately as I do about these plantigrades—mammals that walk on the soles of their feet, like humans. These people have nourished me with the fruit of their knowledge and experience, often gathered right at the source, up close with the animals. They've opened hidden doors for me, revealing secret passageways that have led me to gain a greater understanding and enabled me to see a broad cross section of this animal family.

What truly fascinates me, beyond the sheer elegance of the polar bear and the so-called friendly nature of the brown bear, is how ever-present these magnificent animals are around the world, in our various cultures, in the news, and in our imaginations.

I set out to walk in the footsteps of bears at a time when my fellow Frenchmen were hardly interested in polar bears at all. I rose to the challenge to raise awareness about the species that in a few short years had become a symbolic figurehead of climate change. Following in the footsteps of the white bear naturally led me closer to the brown bear. My research was like a time machine, helping me trace their evolution. I've collected many memories along the way, each one finding a place in my mind and resonating with the others. Far from erasing the memory of earlier experiences, every new observation nourishes, enriches, and completes the previous ones, like an endless quest. Similarly, every observer of bears adds to our knowledge, enhances the overall quality of our observations, and strengthens the unique relationship that brings us together. A great many books, travel journals, and scientific studies have been written about bears, and reading and rereading these is an important part of the quest for knowledge, which is only truly meaningful when it is shared. In all my films, books, and talks, I try to paint a complete picture of these great carnivores so that people can appreciate the full extent of their depth and diversity rather than idolizing them or seeing them as a means to an end, which happens all too often.

I've had the privilege—and it is indeed a privilege—to see hundreds of polar bears, from Baffin Island to northern Alaska, to spend no fewer than twenty-three fall seasons in Churchill, Manitoba, to be one of the first to photograph bears emerging from their dens, and to film polar bears fishing in the Labrador Peninsula. I've been fortunate to observe and learn about brown bears not only in Northern Japan, Siberia, and Russia's Kamchatka Peninsula, but also in Alaska, British Columbia, and Finland. By observing bears for hours, days, and years on end, I've come to a deeper understanding of the conviction shared by those in many cultures around the world who have been close enough to these creatures, for long enough: that bears are a reflection of ourselves. They reflect the wild side of human nature, the way the first humans lived, as they once shared our habitat and stirred our imaginations. Believe it or not, humans and bears once lived in harmony. For thousands of years, bears have nourished the legends, customs, and beliefs that remain deeply rooted in our human consciousness to this day.

Yet at some point, things came to a head and we went our separate ways. Just as Cain killed Abel, and pioneering farmers put an end to the ways of nomadic shepherds, humans and bears, two beings cut from the same cloth, set about tearing each other apart. Humans emerged as the victors from this fratricidal battle with bears. But by destroying bears' habitats, by hunting them intensively, we humans have been the authors of our own misfortune, distancing ourselves from nature. Perhaps there's a lesson for us here. Let's not forget, in both the Bible and the Qur'an, Cain is cursed and his descendants perish in the deluge. Look out: the floodwaters are quickly rising.

CHAPTER 1

How to Describe a Bear

Then he saw the bear. It did not emerge, appear:
it was just there, immobile, fixed in the green and windless
noon's hot dappling, not as big as he had dreamed it
but as big as he had expected, bigger, dimensionless
against the dappled obscurity, looking at him.

William Faulkner[1]

SEEING A BEAR pass by in the forest, slinking between the birch trees and the alders, is one of the greatest moments of serenity I can imagine. Every day I look at a color engraving by Robert Hainard hanging on my wall that takes me right back to moments like these. Emerging from the birch trees, the bear stands observing the observer from afar. This work of art captures it all, conveying a hint of mystery, washing the scene with a gentle light that fades into a pool of serenity and moves the observer profoundly. It's like an elixir of youth, giving you a taste of humility and a sense of harmony.

How do you describe a bear? In my opinion, nothing captures the difficulty of this task better than the opening sentence to the section entitled "The Bear" in *Buffon's Natural History*, from 1797:

> There is no animal so generally known, about which naturalists have differed so much as the Bear, their doubts and even contradictions, with respect to the nature and manners of this animal, seem to have arisen from their not distinguishing the different species, and consequently ascribing to one the properties belonging to another.[2]

Bears come in many different shapes and sizes, as these lines written over two hundred years ago sum up so well.

Popular imagery likes to paint a picture of bears moving around on their hind legs, like furry forest-dwelling people—admittedly a little fragrant, but somewhat good-natured and as family-oriented as humans. Few of us are fortunate to observe bears in their habitat, but we all seem to have our own ideas and ideals about their nature. The stories we tell our children, and the tenderness with which they hug their teddy bears in bed at night, tend to make us forget how complex a species—and how big a family—bears really are.

It always comes as a surprise to see a bear appear in the forest, because it emerges with no sound to announce its arrival. There's no beating around the bush. This is the bear's home, and it is perfectly at ease. As soon as we observers see a bear emerge, these are the questions we ask: Is it a male or a female? How old is it? Is this particular bear known to frequent this area? These are not always easy questions to answer.

A young male may be as svelte as an adult female. Every little detail can provide a clue and help us identify an individual bear. For instance, scars will often suggest that we're dealing with a male bear who has been in a fight or two during mating season, while prominent nipples typically characterize older female bears.

There's something about a brown bear's gait, the way it carries itself. A brown bear tends to roll its shoulders like a sumo wrestler, always ready for combat. Their gait has nothing of the elegance of the polar bear's measured paw plants, but it would be foolish to assume they lack finesse. Bears very rarely even snap a twig as they move through the forest, unless they want to announce their presence. In the bushes, their fur allows them to move stealthily along without ruffling any branches. Unlike felines, with their athletic musculature rippling just below their fur, bears are cloaked in a thick pelt that reveals nothing of their formidable fighting prowess. There are actually three layers to a bear's fur: a short, downy undercoat next to their skin; an outer layer of long, strong guard hairs; and an intermediate layer of awn hair to bridge the gap. The thickness of the hair can vary depending on the season and from one bear to another. A bear's fur serves many purposes, providing protection against not only the ravages of harsh climates but also physical damage from life in the forest—and the occasional swipe of another bear's paw.

What fascinates me the most about brown bears is probably the sheer variation in their color, appearance, behavior, and character. The brown bear is polymorphic, in that there are various subspecies, unlike the polar bear, which is a far more homogeneous animal. (In North America, brown bears that live inland are called grizzly bears; grizzlies are considered a subspecies of brown bear, though the distinctions are somewhat arbitrary.)

Interestingly, the brown bear is perhaps the only species whose binomial name, *Ursus arctos*, is composed of two equivalent terms, the first being the Latin word for "bear" and the second being derived from the Ancient Greek equivalent, *arktos*. Incidentally, the name for the Eurasian brown bear subspecies is *Ursus arctos arctos*. It would be hard to define an animal any more conclusively than that. As if it were necessary to stress a uniqueness this bear doesn't actually possess. *Ursus arctos*, the "bear bear," the real thing. The figure of our imagination, the prototype, the reflection of ourselves.

We can learn a lot from the etymology of the word *bear*. The modern-day term can be traced back to the Indo-European root *bher*—meaning brown—as well as the Old English *bera*, and *beron* in the early Germanic languages. Today, the German word for bear is *Bär* and its Norwegian counterpart is *Bjørn*, also a common male first name. In fact, many names and places have their linguistic origins in the Anglo-Saxon term, such as the first name Bernard and the capital city of Switzerland, Bern.

Another Indo-European root, *rksos*, gave rise to both the Latin word *ursus* and the Ancient Greek *arktos*. It doesn't take much imagination to see the connection between *ursus* and the words for "bear" in many of the Romance languages—*ours* in French, *oso* in Spanish, and *orso* in Italian, for instance.

In the binomial nomenclature of living species, *Ursus* is the name of a genus in the bear family Ursidae, and Ursa Major is the name of the constellation in the northern sky more commonly referred to as the Great Bear. It's no coincidence that the Greek word *arktos* is also used to refer to the north. The adjective *arktikos* not only translates to "arctic," as in "northern," but as professor of ancient literature Charles-Frédéric Schmitzberger[3] points out, it also shares a homonym with the word for "initial," which is reflected in the English words *archeology* and *archetype*. It may be tempting to join the dots and assume that bears are synonymous with "the beginning" in a historical sense, but the etymological roots are not the same. While Charlton T. Lewis and Charles Short's *A Latin Dictionary* (1879) cited the meaning of the word *orsus* as "begun," there is no connection to the word for bear.

Many historical figures owe their names to the Greek *arktos*. One prominent example is King Arthur, whose name is said to come from *Arktouros*, a name that can be translated as "guardian of the bear." Artemis, the Greek goddess of hunting, may also take her name from the bear, though the etymological connection is unclear. It's also worth mentioning that both the Greek *arktos* and the Latin *ursus* stress the first syllable, as if insisting on a power to be both feared and revered.

While linguists have been able to trace the origins of the bear's name with some degree of certainty, its scientific nomenclature has proven more difficult to define. Because of the variability of morphology among bears, scientists have disagreed on how best to name and classify them. The Comte de Buffon, quoted above, described the differences among bears somewhat instinctively, through empirical observation. For Carl Linnaeus—the father of binomial nomenclature, who wrote about the systems of nature in the late eighteenth century—this approach was much too subjective and lacking in strict criteria. "Buffon separates them into carnivorous brown bears and frugivorous black bears," wrote Jean Emmanuel Gilibert, who translated Linnaeus's works into French. "Do all these variations not prove that these distinctions are chimeric?"[4]

Still, the morphological variability Buffon described spurred many attempts to define subspecies, with up to forty being identified. One group described as having dark fur could be isolated here; another group of smaller bears with lighter fur there; a third that were said to be scavengers; or a fourth that preferred to eat ants. An abundance of subspecies emerged, with names just as varied as their characteristics, for example: *albus*, 1788; *alpinus*, 1814; *annulatus*, 1827; *argenteus*, 1827; *aureus*, 1855; *badius*, 1798; *brunneus*, 1827; *cadaverinus*, 1840; *euryrhinus*, 1847; *eversmanni*, 1864; *falciger*, 1836; *formicarius*, 1828; *fuscus*, 1788; *gobiensis*, 1992; *grandis*, 1864; *griseus*, 1792; *longirostris*, 1840; *major*, 1820; *marsicanus*, 1921; *minor*, 1820; *myrmephagus*, 1827; *niger*, 1788; *normalis*, 1864; *norvegicus*, 1829; *polonicus*, 1864; *pyrenaicus*, 1829; *rossicus*, 1864; *rufus*, 1797; *scandinavicus*, 1864; *stenorostris*, 1864; and *ursus*, 1772.

American experts developed one list of subspecies, while their Russian counterparts developed their own. Meanwhile, in Europe, specialists were more hesitant. They realized it was simply impossible to define a hard-and-fast rule as to which coloring was the most dominant. Was it the light brown of the small bears found in Sweden, the charcoal gray of males in Finland, the reddish tone of the magnificent bears on Russia's

Kamchatka Peninsula, or the grayish coat of their Alaskan cousins? I have been fortunate to encounter them all.

Some bears can't even be described by a single coloring. Jean-Jacques Camarra, certainly the leading specialist on the bears of the Pyrenees, offers this poetic description of one of his countless observations:

> Naturally, this is a female, as her poise and finesse suggest. Her coat is simply bicolor, with no gradation between the chestnut brown of her limbs and the beige of her shoulders.[5]

Personally, I find the brown bears of Hokkaido the most astonishing of all. Their ears seem out of proportion to the rest of their body. In spite of their summer diet of salmon, they're not as bulky as their cousins in southern Alaska or on the Kamchatka Peninsula, and their heads are quite narrow. They're generally light brown in color, and many of the individuals I've encountered had a beautiful orange-hued collar-like marking on their necks, the likes of which I've seen nowhere else. Their young also tend to be more slender and longer legged than other cubs. The Hokkaido brown bear has been classified as a distinct subspecies, *Ursus arctos yesoensis*. Living in isolation on the island of Hokkaido, formerly known as Yeso, has been conducive to these bears developing their own physical traits despite the relative proximity of the bear population on Russia's Sakhalin Island to the north.

None of these descriptions carry much scientific weight, as there are no real rules regarding color variation within a bear population; subspecies are defined by genetic differences rather than morphological traits. However, they are a reflection of the fascinating things one can discover by observing these majestic carnivores. As naturalists, we are always looking for what distinguishes one specimen from another; this is what makes individual encounters stand out in our minds and helps us relive and appreciate these magical moments. Given that we seasoned observers have seen dozens or even hundreds of bears, it's impossible for us to remember them all. Some, however, will always stand out, and we'll never tire of seeking new encounters with these magnificent animals.

How genetics can provide clarity

RESEARCHERS HAVE BEGUN to study and classify the genetics of a number of brown bear populations to chart their evolution. Among the populations currently sequenced are several groups in Eastern and Western Europe, North America, and Tibet; three distinct groups on the island of Hokkaido and one on each of the islands of Kunashir and Iturup, to the north; and one group in western Alaska. Interestingly, genetic sequencing has determined that one subspecies of brown bear in southern Alaska's ABC Islands (Admiralty, Baranof, Chichagof), the Sitka brown bear (*Ursus arctos sitkensis*), is related to the polar bear.

Besides differences in coloring, bears can also vary in size, not only from one region to another, but also throughout the year. The geographic variation seen between different bear populations can be explained by their diet, which can vary tremendously in protein content from one region to another.

Within each population, male bears are notably bigger, taller, and heavier than females. Some of the smallest brown bears are found in the Pyrenees, which divide France and Spain, and in the regions of Trentino and Abruzzo in Italy. Based on observations, adult females here weigh around 165 pounds (75 kg), and young adult males around 250 pounds (115 kg), while older males may weigh up to 650 pounds (300 kg). By comparison, in southern Alaska and on the Kamchatka Peninsula, adult females range from around 400 to 700 pounds (180 to 320 kg), while adult males typically range from 600 to 1,400 pounds (270 to 635 kg), with some individuals as large as 1,500 pounds (680 kg). No other species of land mammal exhibits such a great variation in size and weight between the smallest adult female and the largest adult male!

The significant difference in body mass between males and females can be explained by the competition in which males engage with one another during mating season. While the sex ratio of bears is more or less 1:1, females only reproduce every three years, once they've finished raising their cubs. This means that only around a third of females are available for mating each year, so there is stiff competition among males. Before they can pass on their genetic heritage, young males must wait until they're around ten years old before they're strong enough to challenge older males—who may be bigger and tougher but are perhaps losing stamina.

Variations in weight throughout the year are observed in both sexes. It's not unusual to see a bear who was sporting a lean, slender physique in the spring become considerably rounder and stouter by the fall. In fact, a bear's weight may vary by as much as 30 to 50 percent over a twelve-month period, making it difficult to identify an individual by shape alone. That's why, to reliably identify an individual bear, observers look for distinguishing features or markings—such as scars in males, rings around the neck, or patches of contrasting fur color.

What bear tracks can tell us

BEARS MAY NOT always make their presence known, but they do leave tracks. And they can easily travel ten to twelve miles (15 to 20 km) in a night. Even though they walk on all fours, they use their front and hind paws differently, so they leave two types of tracks. In terms of their gait, bears are considered both plantigrade and digitigrade, in that they walk on the soles of their hind paws and the digits of their front paws. When we compare their front and hind paw prints, one looks more like a footprint while the other resembles a handprint. Typically, a bear's hind paw print is about as long as a men's size 10 shoe. Shepherds in the French Pyrenees used to nickname the bear *le va-nu-pieds*—"the barefoot man"—or *lou pé des caous* in the local Béarnese dialect, meaning "the wanderer" or "man of the mountain." Unlike with wolf tracks, for instance, there is no mistaking a bear's tracks for any other wild animal's. What's more, there are distinguishing features that make it possible to identify individual bears by the tracks they leave.

Brown bears—especially those found in Alaska and along the Kamchatka Peninsula—have long claws, sometimes up to four inches (10 cm) long, that leave clear marks on the ground.

A bear's claws are extremely well developed, particularly those on their front paws, which are more often used to dig holes than kill other animals. Bears also use their claws to scratch trees in highly visible places around head-height for humans. They tend to seek out resinous trees in particular because the smell of the turpentine oil excites them. They rub their bodies against the tree trunk not only to mark their territory but also to get rid of parasites. Sometimes we find bits of a bear's fur stuck to the resin on a tree trunk.

Bears also tend to leave their droppings wherever they walk. These droppings—called scats—can be a mine of information for researchers. Because bears are carnivores, their digestive systems are ill-adapted for digesting the plants that nevertheless make up a large part of their diet. This means that part of the food they eat is eliminated either undigested or partially digested. As a result, berries, pine nuts, nutshells, and grasses, as well as fish bones and scales, are often found in their scats.

Thanks to modern DNA investigative analysis techniques, researchers are able to accurately identify the scats they find as belonging to specific individuals, and use these to track the bear's movements and study its dietary habits. Researchers also sample the hairs that bears have left behind on trees or in special hair-collection devices set up along bear travel routes. Together, these non-invasive techniques allow us to gather reliable information about bear populations without disturbing them.

Fresh scats are a sure sign that a bear passed through an area recently. Sometimes the bear might still be close by, as this hair-raising anecdote from bear biologist Dennis Sizemore, recounted in Rick Bass's book *The Lost Grizzlies: A Search for Survivors in the Wilderness of Colorado*,[6] illustrates:

> "This is how I saw my first grizzly in the wild," Dennis says. "I was up on the north fork of the Flathead picking through a scat I'd just found. I happened to look up and saw the bear sitting on the hill above me, just watching me. He looked down at me all puzzled, like 'What are you *doing*?'"

We've looked at how bears travel on land, but it's also important to mention their aptitude for swimming. Anyone who thinks that camping on an island will keep them safe from bears is mistaken. Brown bears are strong swimmers; they use their front paws like dogs. They're able to cross rivers and lakes, swim upstream, and even dive below the water for short amounts of time to catch salmon.

Bear cubs learn to follow their mothers into the water at a very early age. Of course, the water is a fun place for them to play. And do bears ever like to play! Though we should always be wary of simplistic anthropomorphism, many observers have reported seeing bears repeating certain playful actions, such as throwing an object up in the air or sending it sliding across the ice. They do seem to enjoy playing together and being in the water or in the snow. A duo of bears can sometimes spend hours play-fighting together—taking turns jumping on each other, splashing each other, and rolling around in the water. Such rough-and-tumble play must be an enjoyable way to cool off on a hot summer's day in Alaska or on the Kamchatka Peninsula.

A real character

HAVE YOU EVER heard the sayings "he's a bit of a bear" or "he's like a bear with a sore head"? We sometimes use these expressions to describe someone who's not very sociable or communicative, in a bad mood, perhaps, or maybe just a little grumpy. Are bears really like this, though? And if so, is this always a true reflection of their character? If there's one thing that observers, scientists, photographers, and hunters all agree on, it's that bears have their own character—or personality, dare I say.

The French naturalist and author François Merlet sums this up rather poetically:

> The bear is a tough nut to crack; one must use one's imagination to peel away the layers of that thick shell of his. This is the key to revealing the secrets of his true character, appreciating his charming personality, and grasping the extent of the astonishing abilities that more than justify his almighty power.[7]

One of the great American trackers, Doug Peacock, shares some cautionary words on this topic: "It is impossible to predict with certainty the temperament of the bear with whom you are dealing."[8]

Overall, I'd say that bears are a little like humans, in that the majority are generally agreeable, though some can be short-tempered and, albeit rarely, dangerous.

I've often observed differences in character among bears in the same family. Even at a very early age, some cubs are clearly more adventurous while others are more fearful, timid, or aggressive. In spite of a seemingly identical education and upbringing, cubs in the same family can show distinct character traits, just like human siblings.

On several occasions, in Eastern Siberia and Japan, I've seen twin bear cubs exhibit very different reactions and abilities. On Hokkaido, I once observed two young bears whose mother had recently left them to their own devices, now that her job was done. It might have been a few days, or even a few weeks, since they had been on their own; either way, this brother and sister had been sticking closely together. It was during the salmon run, so there was abundant food ready for the taking. One of the young bears, always the most active of the pair, spotted the entrance to a channel the salmon were using to swim up the rapids, and scrambled down to the river. However, she slipped and disappeared completely from her brother's line of sight, and he desperately began to whine, seemingly panicking at the idea of losing his companion and finding himself alone. After a few minutes, the more adventurous cub managed to scramble her way to safety. Her brother bounded straight over toward her, seemingly relieved to see her again, only for her to give him a big whack on the snout with her paw!

This may seem like a very anthropomorphic description, as if I were somehow humanizing their interaction, but it was precisely the way the scene played out. What's more, I've observed this type of behavior on several other occasions. In fact, it's not uncommon among siblings for one of them to habitually beg the other for scraps of salmon, finish off any leftovers, and hope for the occasional gesture of kindness. Going back to the story of the young siblings in Hokkaido, I saw the pair make their way down to the water together, and the more adventurous one caught a salmon, generously gave it to her brother, and went back to catch one for herself.

In his guide to the wild mammals of Europe, Robert Hainard shares his own thoughts:

> Personally, I find it hard to reconcile the varying opinions about bears and their character. The bear is certainly a very cautious creature, to say the least, who takes great care to avoid humans, and is aided in this endeavor by his great intelligence and subtle senses.[9]

Some bears may be irritated by humans in general, or even just by a particular individual. On one memorable expedition, we were navigating our way around the vast Nekite River estuary on the coast of British Columbia, from our base at the Great Bear Lodge. The owner, Tom Rivest, has the greatest of respect for grizzlies and their territory, and he knows better than anyone how to approach and behave among them. No one was allowed to leave the boat and set foot on land, so as not to disturb the bears' habitat. In spite of these precautions, one of the grizzlies we saw would not tolerate Tom's presence. As soon as the boat drew closer to the beach, the grizzly would lower its ears and growl. Perhaps it associated Tom's smell with a bad memory. It was a good thing the boat didn't get stuck, as things could have turned ugly if Tom had had to jump out and push it out of the shallows.

Bears and their traits of character feature prominently in many Indigenous stories and legends. In various First Nations and Native American cultures, bears are portrayed with as many types of character as one can imagine. They may be kind and generous, or they may be mischievous and mean-spirited. There are some bears with noble intentions who would sacrifice their own life to save their family, and others so selfish they hide food and keep it from others. There are those who are cunning, and others who are easily fooled, just as there are bears who are destructive and menacing, and bears who are caring and considerate. As far as stories and legends go, no other beings apart from humans are described with such great diversity of character. Like the Greek gods and goddesses in ancient mythology, bears in the oral traditions of the Indigenous Peoples of North America exhibit all the qualities and shortcomings we usually only attribute to humans.

As we can see, bears can be real characters. Perhaps this is what has enabled them to disperse so extensively into new territories and thrive in extreme habitats.

Before humans spread themselves so widely across the planet, bears could be found all over the Northern Hemisphere. There were not only the brown bears, polar bears, and pandas we're familiar with today, but also species that are now extinct, such as short-faced bears and cave bears. Some were carnivores, while others were omnivores or vegetarians, or ate highly specialized diets.

Brown bears have always been the most widespread around the world. They are masters of adaptability, which allows them to live in greatly varied ecosystems, from the arid environments of the Gobi and Sinai deserts to the vast glacial plains of the Canadian Barren Lands, as well as in all kinds of temperate and boreal forests. They sometimes venture to altitudes as high as 6,500 feet (2,000 m). They climb to great heights, cross mountain passes, and are capable of covering vast distances, even finding their way across rivers and lakes. Some brown bears—mostly males—are known to have traveled extremely far north, even venturing onto the pack ice.

Only one other species has succeeded in thriving in as many different habitats: *Homo sapiens*.

Bears in the art world

Bears are rarely seen in the great paintings of the world. As Michel Pastoureau suggests in his book *The Bear: History of a Fallen King*, the bear was once venerated but later dethroned as king of beasts by the lion, thereafter making only rare appearances on canvas.[10] However, the bear features prominently in the artwork of the *Livre de chasse*, one of the earliest books on hunting, written by Gaston Phoebus in the late fourteenth century.

In my mind, the finest work depicting a bear is *Morning in a Pine Forest*, credited to Russian artist Ivan Shishkin (1832–98) but created in collaboration with his peer Konstantin Savitsky (1844–1905). Shishkin painted the bucolic scene showing a family of bears in the forest with a naturalistic vision and realism worthy of the Barbizon school of painters.

Other painters have also created works of great interest. For starters, the French artist Philippe Legendre-Kvater has introduced two generations of children to simple and effective drawing techniques. His little how-to books teach children how to draw a bear, starting with a big circle for the head, a smaller circle for the nose, an even smaller circle for the snout, two circles inside one another for the eyes, two more little ones for the ears, and a semicircle for the mouth... and there you have it! It may be a simplistic, stylized bear, but it's still a bear. In his books, Philippe also tells children stories to educate them about bears and the bear tamers of the past.

I've always been impressed by wildlife painter Éric Alibert—with whom I've been fortunate to share several bear observation expeditions—and the remarkable speed with which he is able to draw bears in the wild. I can still clearly picture him by the side of a river in Eastern Siberia, one eye on his binoculars, the other on his sketch pad, painting the bear grazing on the other shore with three swirls of his watercolor brush. In just a few strokes, he brought that bear to life, smoothing out all the rough edges from its fur. I also remember the day I saw Éric pick up a copper plate and a stylus on the fly to engrave an image he would later turn into a series of prints. A few simple overlapping, intersecting lines, and there was the bear, just like in the thirty-thousand-year-old paintings of bears found in France's famous Chauvet Cave. Depending on the light and the angle from which you look at this work, the bear seems to move as if by magic, like an early rendering of an animated movie.

Meanwhile, Robert Hainard (1906–99) depicted bears in a wholly different style, representing a different era. Often, he saw the bear as part dog, part wolf, a dusky silhouette rendered perfectly by his technique of engraving on wood. The works he sketched by moonlight, capturing the contours of his subjects, are almost surreal, ghostly observations, yet the bears are right there, clear as day.

Sculpture tells another story entirely. Among the great French sculptors, perhaps one of the best known was Emmanuel Frémiet (1824–1910). His sculpture *Le Dénicheur d'oursons (The Bear Cub Thief)* stands in the Jardin des plantes in Paris as a towering example of our romantic vision of this majestic animal. Depicting a bear baring its teeth menacingly at a hunter making off with a bear cub, this work is so realistic that the bear's fur seems to be rippling, as if there were muscles beneath its bronze pelt.

François Pompon (1855–1933) is known for his pure, minimalistic style, which renders only the essential. His stone sculpture of a polar bear in stride, which stands in the Musée d'Orsay in Paris, may have brought him widespread recognition and fame, but his smooth, rounded rendering of a brown bear is just as beautifully crafted, albeit not quite as spectacular.

Michel Bassompierre is a shining example of France's contemporary animal sculptors. His understanding of how bears move, how they lift their paws, and how their hair rounds their silhouette, is meticulous and simply remarkable. His sculptures seem to move before your eyes, their every voluptuous curve and subtle shadow hinting at the anatomy beneath the surface and making these bears larger than life as they turn to look at you inquisitively.[11]

One of Canada's most notable contemporary Indigenous visual artists is Jason Carter, a member of the Little Red River Cree Nation in northern Alberta. His passion for nature—and bears in particular—is clear to see in his bold, original stone sculptures and colorful artwork depicting grizzlies sitting, standing, sleeping, and rolling their shoulders. In his work, he captures a beauty that transcends the power of these remarkable creatures.

Rendering a bear in a work of art is a way of seeking one's reflection, one's alter ego, one's twin. One must look beneath the surface, beyond the rolling curves of its body, to uncover the wildness that lies within.

CHAPTER 2

How to Become a Bear

It looked at me as tenderly as a bear could well do with one eye; it opened its mouth, not in ferocity, but yawningly. This bear had something of peace, of resignation, and of drowsiness; and I found a likeness in its physiognomy to those old stagers that listen to tragedies. In fact, its countenance pleased me so much that I resolved to put as good a face upon the matter as I could.

Victor Hugo[1]

IT'S TIME FOR a little story about brown bears and how they and other species of bears evolved. Once upon a time, about 57 million years ago, in the forests of Central Asia, there was a family of small carnivorous tree dwellers, catlike creatures that bore some resemblance to today's martens and civets. The members of this family are known as miacids and are said to have been at the origin of the various groups of carnivores we know today.

This brings us to the next chapter in the story, which takes place some 42 million years ago. Around this time, a branch developed that would lead to the evolution of three related families: the Otariidae (pinnipeds such as seals and sea lions), Mustelidae (carnivorous mammals such as weasels, otters, and wolverines), and Ursidae (bears). It might seem like a bit of a stretch to compare sea lions to bears, but the secret is in their skulls. If you were to hold the skull of a South American sea lion (*Otaria flavescens* or *Otaria byronia*) in one hand and the skull of a brown bear in the other, the similarities would be clear to see. Presumably this marine mammal evolved from a close relative of the bears that adapted to life in a watery environment. The animals also have similar teeth, but while the brown bear has developed some omnivorous habits, the South American sea lion is still very much a carnivore, feeding mostly on fish and, in some cases, penguins.

On Steeple Jason Island, which lies to the northwest of the Falkland Islands, I once had the opportunity to observe a South American sea lion hunting for penguins, catching them as they were going out or returning from fishing. It would first grab a penguin with its powerful canines and then tear it apart by shaking and pounding it against the surface of the water to remove its thick, feathered skin. I saw this sea lion snatch a large fish the same way with its teeth, a skipjack tuna that must have weighed about twenty pounds (10 kg), but this time it tilted its head back and swallowed the fish whole, the way a reptile would. This marine mammal's teeth are ill adapted to chew and break up the flesh of its prey, even the tender meat of a tuna fish.

Now if we compare the skull of a brown bear with the skull of its polar bear cousin, their teeth can tell us a lot about how these two species evolved. Generally speaking, species with a specialized diet tend to become increasingly opportunistic—eating a variety of foods—but the polar bear has bucked the trend as it has evolved. Its diet has become highly specialized, and its teeth are more like those of its pure-carnivore ancestors.

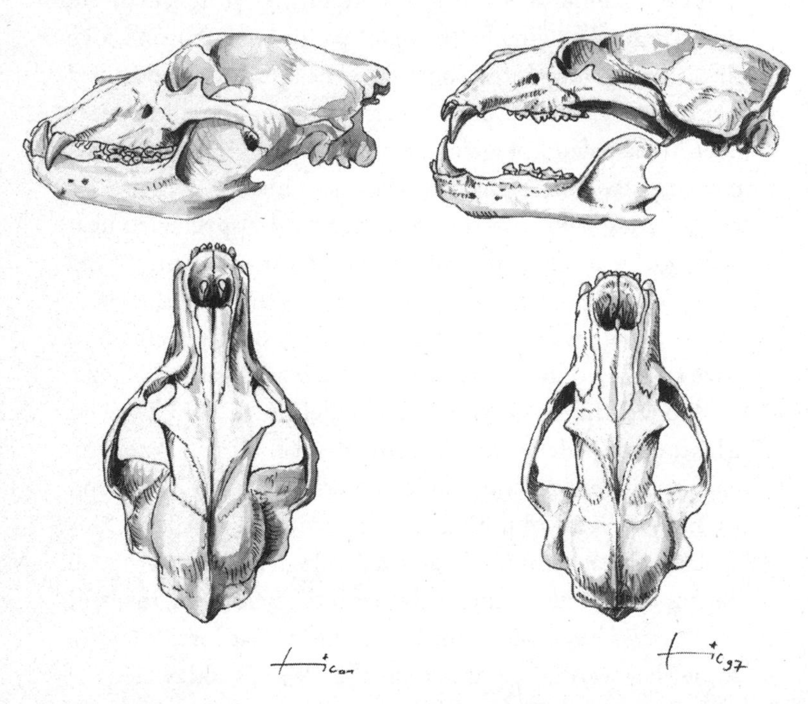

Comparison of the skull of a brown bear (left) with the skull of a polar bear (right)

Meanwhile, the teeth of a brown bear are those of an omnivore, resembling a wild boar's teeth, for instance: large canines to seize its prey and rounded molars with no prominent cusps that are well suited for chewing plant roots.

The Ursidae family is relatively new in evolutionary terms, dating back to some 25 million years ago. *Ursavus elmensis,* a long-extinct fox-sized carnivoran mammal also known as the dawn bear, was the first member of the family. Relatively quickly (15 million years ago) three further groups evolved: one eventually giving us the giant panda we know and love, the second giving us the short-faced bear branch, and a third group that led to all other bears. Another 5 million years later in the paleontological layers of Western Europe, we discover *Ursus minimus,* the direct ancestor of all known species of modern bears, with the exception of the spectacled bear (*Tremarctos ornatus,* the last remaining descendent of the short-faced bear branch). Also known as the Auvergne bear, *Ursus minimus* was small, as its name suggests, and thought to have resembled the modern American black bear.

Around the seventy-eighth parallel north on Ellesmere Island, the skeletal remains of a small bear (*Protarctos abstrusus*) were recently collected and found to be around 3.5 million years old. There's evidence to suggest that this species of bear must have eaten a diet high in the sugar-laden berries commonly found at this latitude. Back then, the landscape must have been similar to the boreal forest, with a somewhat warmer climate than the area has today, but the same long nights. This was likely the first bear to emerge from Asia around 5 million years ago and make its way toward the

American continent. Various closely related species have been discovered in China, Eastern Europe, and the United States.

Around the turning point between the Pliocene and Pleistocene epochs, some 2.1 million years ago, the family diversified and dispersed. In Europe, the Etruscan bear (*Ursus etruscus*) is thought to have been the direct descendant of *Ursus minimus*. The Etruscan bear also gave rise to some extinct species, such as the cave bear (*Ursus spelaeus*) and its ancestor, known as Deninger's bear (*Ursus deningeri*).

The cave bear was essentially vegetarian, albeit a hefty beast. The average weight of a female cave bear is estimated to have been around 500 pounds (225 kg), while her male counterpart is thought to have tipped the scales at closer to 925 pounds (420 kg). If you're ever in France, the best way to get a sense of what this bear looked like is to visit Chauvet–Pont d'Arc, where you can find a replica of the original Chauvet Cave. The paintings you'll see on the walls of this cave are strikingly realistic and seem to come to life in the shadows. The cave bear clearly had a strong head, though its jaws don't seem to have been as powerful as a brown bear's. Its muzzle was quite pronounced, and it had the broad shoulders of an animal that was used to digging in the ground.

The cave bear was still alive at the time when the modern brown bear began to colonize Eurasia, but it left no descendants when it succumbed to the human invasion some ten to twelve thousand years ago. One thing I'd like to point out here is that paleontologists have often found signs of bone disease when examining fossils of these bears. It would be interesting to compare these findings with the current research on the brown bear. The cave bear may have disappeared around the time humans came onto the scene, but it seems unlikely that our ancestors were the only reason for its demise. Perhaps it simply wasn't able to adapt and develop the ability to hibernate over the increasingly long winters during the last glacial period. Now that I've planted that seed, I'll let it sit for a while before coming back to explain more in chapter 8.

The short-faced bear (*Arctodus simus*) that once lived on the plains of North America was also a contemporary of the brown bear, but found itself in direct competition with its human rival, as both *Homo sapiens* and *Arctodus simus* hunted the same large herbivores. This huge bear—a large male weighing in around two thousand pounds (900 kg)—had the paws of a long-distance runner. As it roamed the great wide open, it likely confined the black bear to the forest and forced the brown bear to stay in Alaska rather than migrating southward. Try to picture the scene, with millions of bison and pronghorns—the beige-and-white antelope-like creatures indigenous to interior western and central North America—grazing on the vast grassy plains. Imagine a saber-toothed tiger crouching in wait, ready to pounce on its prey, when suddenly a great big bear with a head not too different from the modern bear's emerges from out of nowhere. This bear stood over eleven feet (3.3 m) tall on its hind legs—that's a head or two higher than a polar bear. With the muscular hindquarters of an animal built to run, it must have been a mighty predator.

Today, its sole remaining close relative, the spectacled bear (*Tremarctos ornatus*), is the only species of bear found in South America—in the mountains of Peru, Venezuela, and Colombia.

The earliest fossils that have been attributed to a direct ancestor or the modern brown bear are 1.2 million years old. All the paleontological and genetic evidence supports the theory that the first brown bears evolved in Central Asia, from where they went on to populate Eurasia and North America, taking advantage of the vast glacial plain that once bridged the gap between the two continents. The brown bears of the Pleistocene epoch were bigger and bulkier than those we see today, and, as their teeth suggest, they were more carnivore than omnivore. The brown bear only started to move into the plains of America after the demise of the short-faced bear some ten to twelve thousand years ago.

Evidence of three close relatives of the Asiatic black bear (*Ursus thibetanus*) has been found in the Var region of Southeast France, and other bone fragments have come to light in Italy and Belgium. These discoveries illustrate how the three different species were dispersed across Southern Europe during the Middle Pleistocene era (786,000 to 130,000 years ago) and suggest a certain geographic isolation during this period, in which there were three successive major glaciations. Another important thing to point out about these discoveries is that during this period, at least five species of bears coexisted in Europe, in addition to the polar bear, which was found on the periphery of glaciated areas.

Many mammals—canids such as wolves, African wild dogs, hyenas, and foxes, as well as mustelids such as badgers—use dens to protect their young. Brown bears are no exception, in spite of their size. These large mammals likely developed this unique behavior as a result of climate change in the transition between the Pliocene and Pleistocene eras. With temperatures dropping and glaciation becoming more frequent, conditions were favorable for species that were able to weather increasingly longer winters and go without food for months on end.

The polar bears that hibernate on the coast of Hudson Bay and James Bay use dens dug out of the permafrost at the tree line. Natural caves are rare in areas like these, but cavities that are made in the permanently frozen ground can last for decades, even centuries. These dens can also provide shelter from the summer heat. Believe it or not, a large part of these areas lie at similar degrees of latitude to England and Scotland, and summer temperatures can sometimes exceed seventy degrees Fahrenheit (21°C).

The family relationship between brown bears and polar bears

THE POLAR BEAR is the more recent of the two species, and is thought to have evolved from the brown bear a mere 600,000 years ago as it specialized in hunting on the pack ice. However, the differences between the two species are not particularly clear-cut, and there are various theories as to how they grew apart, some more far-fetched than others.

At the French National Museum of Natural History in Paris, researcher Alexandre Hassanin has studied the evolution of the brown bear and the polar bear and the influence of glaciations on their diversity. The latest research published on the topic suggests that both species of bears often existed alongside each other, and that the lines differentiating the two are very fine.

With each new glaciation, brown bears moved south in response to the advances of the polar cap, or found themselves trapped in glacial refuges. The glacial valley of Allt nan Uamh in the Northwest Highlands of Scotland is an obvious example. In this deep U-shaped basin lined with rock on either side, at the foot of a rocky outcrop that was spared from the erosion of the huge glacier, lie four impressive caves by the name of Creag nan Uamh. Here, speleologists and scientists have discovered the 45,000-year-old skull of a brown bear, the 22,000-year-old skull of a polar bear, and the 14,000-year-old skull of another brown bear. Along with these bear skulls, the skeletal remains of reindeer, arctic foxes, and wolves have been discovered, as well as a piece of approximately two-thousand-year-old walrus ivory—a clear sign that humans passed through this remote mountainous area of Scotland.

Dr. Andrew Kitchener specializes in research on carnivorans and hybridization in mammals.[2] I met with him in his laboratory at the National Museum of Scotland in Edinburgh, where he spoke to the significance of these caves due to the alternation of the two species of bears in one place. Polar bears were present here at the peak of glaciation and made way for brown bears as the polar caps retreated.

Today, brown bears and polar bears can occasionally be observed alongside one another, particularly in northern Alaska. In September 2015, when I was filming my documentary *Evolution of the Polar Bear,* I wanted to be sure to get footage of them side by side, so we headed to the remote community of Kaktovik.[3] The Iñupiat People here are entitled to hunt three bowhead whales each year when their pods pass by in September. The whale carcasses are then piled in the same place on the beach, which invariably attracts all the bears in the area. We tried our luck, knowing it was a gamble because there were lots of factors that could conspire against our mission. There might be no hunt, if the whales didn't come. The tide might prevent us from approaching the bone pile, or the weather might be too bad to film. Plus, you never know if you're going to find some rowdy locals roaring around on ATVs and scaring the bears away, or if the only grizzly in the area might just have been killed.

When all the stars are finally aligned and the conditions are right, then the waiting game begins, and you have to be extremely patient. This time, we got lucky. It had been dark for a while already, and we could see the ghostly silhouettes of polar bears moving in the night around an enormous fresh whale carcass. It was quite the Pantagruelian banquet for the bears. The only sound amid this peaceful community of Arctic giants was the chomping of their jaws on the chewy whale blubber. In situations like this, the males tolerate even the cubs, and they can be observed feeding almost shoulder to shoulder alongside one another. The mist was fogging up my glasses and the lens of the camera.

Imagine our surprise when the seventeen polar bears we were observing as they enjoyed a quiet dinner barely a hundred feet (30 m) away suddenly scattered like a flock of pigeons in the park and charged right toward us! There was a grizzly in their midst, barging around like, well, a bear with a sore head. The white giants who had seemed so unafraid of anything were certainly in a hurry to leave, even though this wasn't a very big grizzly. The brown bears in this part of Alaska tend to be fairly small, given the limited resources. This grizzly was hungry, and he wanted his pound of flesh too. He was in a hurry as well. Winter was coming, and he would soon be heading for the mountains of the Brooks Range to hibernate. It was an incredible stroke of luck to see this brown bear barge onto the scene, because while the Iñupiat are not allowed to hunt the polar bear, they tend to be trigger-happy when it comes to the brown bear, an animal they see as an intruder to their community and their culture.

At the time we made these observations, only one grizzly was known to be in the area. In spite of the bright light we needed for filming, he gained confidence and returned five more times, spending longer and taking more food with each visit. The polar bears came back too, once they had identified and accepted the newcomer. It was a rare sight to witness the large males sharing a meal like this. Thanks to the traditional hunting activities of the Iñupiat People, this brown bear and his polar bear cousins were able to fill their bellies for a comfortable winter.

It isn't hard to imagine how the interglacial periods of the past must have affected these two species—so close in their genetic makeup, but so distant from each other in their feeding habits.

I couldn't help but wonder why these large white bears were so afraid of the smaller brown bear. Susan Miller studies polar bears in this region as part of her work as a fish and wildlife biologist with the US Fish and Wildlife Service. She explained that September is a crucial time of the year for brown bears, who must eat as much as they can before hunkering down for the winter in their dens. Meanwhile, polar bears take a more pragmatic approach. For them, winter is when they get to roam the pack ice and hunt seals, so they would rather avoid conflict and move aside for a while to let a brown bear have his fill.

The brown bear we saw was understandably aggressive and in a hurry, while the polar bears had all the time in the world and were not interested in picking a fight. But there may be other theories to explain the dominance of the brown bear in this case. In this environment, the brown bear has evolved in competition with wolves, other bears, and wolverines, so it has been forced to defend itself far more than a polar bear living alone in its own food niche would. What's more, for a strict carnivore like the polar bear, even a slight injury could prove fatal, as this would prevent it from hunting.

There is genetic evidence to suggest that female brown bears in the ABC Islands of southern Alaska have mated with male polar bears, enabling their descendants to better adapt to a changing climate. Sequencing of the genomes of both species has revealed a certain mitochondrial introgression. In other words, polar bears still have a little brown bear DNA—about 9 percent, estimates suggest.

Recently, male brown bears have often been observed venturing far to the north and mating with female polar bears. The result of this interbreeding of the two species is a hybrid known as a pizzly or grolar bear, a portmanteau of both names.

So far, only a few have been observed on Canada's remote Victoria Island, but perhaps there are others elsewhere. One reason their numbers are low is that they're hunted by the Inuit, who don't have the right to hunt grizzlies in the area. That raises the question: Is a pizzly brown or white?

However we classify it, a hybrid cub with a grizzly for a father and a polar bear for a mother will grow up learning how to be a polar bear. This cub may not be as physically well-equipped for life on the pack ice, but he will be better able to make it through longer, warmer summers. In terms of evolution, hybridization is certainly a rapid way for bears to adapt to change.

While the color of a pizzly bear's fur is the result of hybridization, another factor is at play in the coloring of the Kermode bear (*Ursus americanus kermodei*), a subspecies of the American black bear living in the Central and North Coast regions of British Columbia. Some Kermode bears are born with white- or cream-colored fur—but these rare creatures are black bears with white fur, not albinos. When we see polar bears walking on the pack ice, or brown bears standing in a forest of tall cedars, it's obvious that their coloring is designed to help them blend in to their environment. Their success as predators—and their survival—depends on it. So why are some black bears born with white fur? There is a rational explanation: a rare recessive gene. Still, it's hard to fathom how two cubs in the same litter can have different-colored fur—one black, and one white.

This rare white-coated bear, known as the "spirit bear," holds a special place in the oral tradition and culture of the coastal First Nations in British Columbia. This is a creature to be respected, not hunted—a reminder of a time when the earth was covered by snow and ice, a relic of the last ice age during which the First Peoples' ancestors traveled vast distances to discover a new continent and establish their traditional territory on the Pacific coast. For these Indigenous Peoples, the spirit bear is a symbol of harmony and serenity and a living memory of generations past; thus, its habitat should be protected so that its wisdom can continue to guide the way in an ever-changing world.

Bear species around the world

BEARS FIRST CAME to the American continent from Eurasia around 70,000 years ago. The current geographic distribution of brown bears is directly connected to the effects of the last glaciation, which reached its peak some 22,000 years ago, forcing all brown bear populations to move south.

Later, as the polar caps rapidly retreated, bears returned to more northerly latitudes to populate the new-grown forests. They were likely the most geographically widespread around 6,000 years ago, when their territory extended across the entire western part of North America and even into Mexico; they could be found all across Europe, with some population dispersal into North Africa, as well as across most of Russia and Turkey, in the mountains of Lebanon and Syria, throughout the Himalayan arc, and all the way up to the north of Japan.

Since this conquest, human activity, urbanization, the destruction of forests and plains, and intensive hunting have fragmented the European population, wiped out the last bears in North Africa, and isolated the few remaining bears in Syria near the border with Lebanon, where they've managed to survive against all odds in the middle of a war zone.

Genetic analysis has highlighted the isolation of bear populations as a result of the fragmentation of their habitat. In mainland Europe, several small groups have found themselves isolated in the Pyrenees, as well as in the Asturias region of northwest Spain, and some bears can still be found farther east, in Slovakia, Romania, and Bulgaria. The last remaining bear in the French Alps was killed in 1921, and the last reported sighting in the Vercors foothills, to the west, was in 1937. The brown bear was wiped out from the British Isles around the year 1000 as a result of intensive hunting.

The Marsican (or Apennine) brown bear, designated by the subspecies *Ursus arctos marsicanus,* has managed to survive in the Abruzzo region and the Apennine Mountains of central Italy. Considered to be quite tolerant of humans, this bear branched out from its Alpine cousins some 4,600 years ago and is smaller than its closest living relatives, in the Balkans. Various factors, including cranial measurements, have clearly shown that this bear constitutes a separate population living in isolation. It also has a low genetic variability and a high degree of consanguinity (inbreeding), both symptomatic of populations deprived of external contact. Meanwhile, in Sweden, two formerly separate bear populations are converging once again, and other groups of bears appear to be trying to return to Switzerland and Austria.

It's one thing to be able to determine the range of places where bears once lived during the Neolithic age, but one crucial piece of information is missing: How many bears were there? Once again, genetics may eventually help us to estimate the size of the bear population using DNA trackers and a sufficiently large sample size and range, but for now, we simply don't know. One clue, however, may lie in the omnipresence of bears in various traditional cultures. That suggests they must have lived in close proximity to humans and been present to some degree in our daily lives—or perhaps they were such a rarity that they led our imaginations to run wild.

Inside the bear's den

WHEN FALL BRINGS the first snow and the biting cold starts to set in, pregnant females seek out a den for the winter, just as other brown bears do, but they tend to look for a quieter, more remote location. After feasting on the last available berries to gain as much weight as they can, they settle comfortably into their dens to spend seven long months isolated from the world and the passing of time outside.

In early January, they give birth to two, three, sometimes even four hairless, blind cubs, each weighing no more than a pound, give or take a few ounces (about 300 to 500 g). That's about as much as three big apples. This might seem completely out of proportion with the size of their mother—it's the equivalent of a human giving birth to a baby the size of a hamster—but there's a logical explanation. After mating takes place in late May or early June, the development of the embryos is interrupted to ensure that the timing of the cubs' birth is synchronized with their mother's hibernation. The implantation of the embryos in the uterine wall is delayed until October or November, when the female retires to her den for the winter. This means that the effective gestation time will only be around fifty days—less than a fifth of the human gestation period.

When her cubs are born in January, the mother will likely have been fasting since October. She must therefore make the best use of her reserves to ensure her own survival and feed her cubs, and she does this by bringing them into the world as early as possible so that she can begin to nurse them. Transforming fatty acids into energy through lactation uses less oxygen for the mother and promotes better growth for the cubs than if they were to continue to develop *in utero*. The cubs are born in a state where they are just able to migrate into their mother's fur and to suck abundantly.

This paradox—an animal the size of a bear giving birth to such tiny, unformed creatures—is thought to have fed into the traditional belief that mother bears lick their young to literally shape them and ready them for life in the outside world. As Pliny the Elder wrote:

> Newborn cubs are a shapeless lump of white flesh, with no eyes or hair, though the claws are visible. The mother bear gradually licks her cubs into their proper shape.[4]

Similarly, in his sixteenth-century opus *Pantagruel,* François Rabelais wrote:

> When it is born [a bear cub is] simply a crude shapeless lump of flesh: then the mother-bear licks its limbs into perfect shape.[5]

By this token, if the mother bear fails to do her job properly, her offspring will have shortcomings. Interestingly, in French there is a term for someone who is gruff and bad-tempered, *un ours mal léché*, or "a poorly licked bear."

As anthropologist Sophie Bobbé has explained, female bears are synonymous with the giving of life. Genesis, seasonal reincarnation, and self-sacrifice through hunting are all attributes that humans have associated with females in their vision of bears throughout the ages. A mother bear's milk is rich in fat—typically around 32 percent. By way of comparison, cow's milk and human breast milk only contain 3 to 5 percent fat. All mother bears nurse their cubs for at least five months, and some may continue nursing for up to eighteen months. Typically, young bears will start to develop a taste for new foods once they leave the den for the first time, at around five months of age.

As they develop in the safety of the den, the cubs grow their coat of hair and their eyes gradually open in the darkness, though they will still have to wait a while before they see the light of day. Once they reach around eleven to thirteen pounds in size (5 to 6 kg), they are ready to venture into the world outside.

Michel Tonelli is a film director who has produced a number of documentaries about bears in the Pyrenees, including footage from cameras placed inside bears' dens. Let's picture the scene. The snow is starting to melt, and the snowdrops that bears love so much are beginning to emerge, drawn out from the earth by the spring sunshine. Inside the den, barely a ray of sunlight has penetrated the darkness, but things are beginning to stir. It's time to venture outside.

The young bear cubs have changed a lot since they were born. These playful little bundles of fluff are now padding around like teddy bears on four endearingly uncoordinated paws, gazing in wonder at everything around them. It's hard to believe such a delicate little thing is going to grow into a giant weighing hundreds of pounds. Their mother has changed a lot over the winter too. She has lost at least 30 percent of her body weight, but she won't be in too much of a hurry to fill her belly. It's time for the new cubs to learn their first lessons in life, so they'd better listen up and pay attention. While bears generally tend to be quiet creatures, mother bears do make subtle huffing and hissing sounds when they're telling their cubs what to do.[6]

The cub's journey to independence

IT'S A RARE and intimate privilege to watch a mother bear nurse her cubs. The young bears are so enthusiastic, and their mother is so protective and maternal, it's hard not to see something of human mothers and babies in their relationship.

When the bear cubs start to get hungry, they make sure their mom knows, by sneaking under her belly and trying to grab a nipple. If it's not the right time, the mother bear swats them away or just ignores them. If it is a good time, she'll circle around and find a comfortable place to settle down, just the way a human mother would to nurse her baby. The difference is, a mother bear has six nipples to feed her hungry young. The cubs jostle with each other for the best spot, and when they've latched on, they start to purr and their mother lays her head back. Sometimes she'll even recline on her back and stretch out long, as if she were enjoying a spot of relaxation. It's a delightful moment to be savored. At least, that's the way it looks to a human observer for those few short minutes.

On Japan's Nemuro Peninsula, my colleagues and I observed a female completely absorbed in the task of fishing for salmon in the river, while her two cubs stood on the shore. Impatient, they grew tired of waiting and ran off a good distance away to play in the woods. Imagine their mother's surprise and concern when she realized that her cubs were nowhere to be seen. In a heartbeat, she was inches away from us with her muzzle in the air, clearly distraught but not showing the slightest aggression toward us. She soon caught the scent of her cubs and bounded off to fetch them. The little ones were hungry, and she started to nurse them right away. Now it was time for them to reconnect after more than a little excitement.

During their two-year upbringing, the young cubs have a lot to learn. They must learn how to recognize the right plants and figure out how to sense and avoid other bears, and they must develop these abilities quickly. Rough-and-tumble is the name of the game for young bear cub siblings. With boundless energy, they climb on tree stumps and shake bushes just for fun, but they never stray too far from their mother. At the slightest sign of danger, they learn to simply climb the nearest tree.

I recall one time when I was at Anan Creek in Alaska, observing black bears in a truly unique environment as they fished for salmon in a raging torrent that flows straight from the rainforest into the ocean. One female seemed to be having a lot of trouble with her cub. The slightest sound out of place would frighten him and send him scurrying up to the top of a tree, about fifty feet (15 m) in the air. The only problem was, every time he went up there, he was too frightened to come down on his own, so his mother would have to climb up too to persuade him that it was safe to slide down the trunk. Once, twice, three times, ten times I counted as the mother bear patiently climbed the tree again to coax her cub down to the ground, whimpering as he came. Female brown bears do the same, in spite of their often imposing size. They'll guide their cubs along just as human parents help their children overcome obstacles.

On another occasion, it was around the middle of September on the Kamchatka Peninsula, and the tail end of a typhoon had blown in from northern Japan. It was raining sideways and the wind had whipped Kurile Lake into a foaming, frenzied sea. Tree trunks were thundering down the river where we were observing the bears, and sandbanks were being washed away by the muddy torrent. Still, a female and her two young cubs, who were just a few months old, decided to attempt a crossing. Mama led the way, with the water coming up to her shoulder as she battled against the white water and the current. The two cubs did their best to follow, but they soon went under. Occasionally, the top of a head and two ears would poke above the surface a little farther downstream. Clearly, they were determined to make it across, but they were scared, as their whimpering conveyed every time they came up for air. Meanwhile, their mother was pacing up and down on the other side, making sounds of encouragement. Eventually, the cubs made it to the other side. They were soaking wet, of course, and simply rolled around in the sand to dry themselves off.

It's not unusual for mother bears to push their cubs outside their comfort zone and use tough love to teach them the ways of the world. When they see their cubs are in difficulty, they'll certainly encourage, guide, and reassure them, though sometimes it's hard to tell which is which. Mama leads by example, and her cubs have to follow as best they can. One cub I observed was always falling behind and ended up going the wrong way. When he realized his mistake, he had to turn around and retrace his steps, because his mother wasn't going to come back and fetch him.

These first two formative years are when a cub learns how to be a bear. Mothers teach their young not only how to look after themselves but also how to live alongside other bears and coexist with other animals. It can be a dangerous time, too, because cubs risk their lives if they don't listen to their mother's advice. Often, female brown bears can be observed emerging from their dens with three cubs, and only two the following year, and sometimes just one will live long enough to be emancipated by its mother and sent out into the world on its own.

Females understandably have a strong attachment to their young, and as the Indigenous Peoples of North America and the hunters of Siberia well know, little can be more dangerous than finding oneself between a mother bear and her cubs. When my colleagues and I were filming a documentary about the relationship between bears and humans in Japan—on Hokkaido as well as on the main island, Honshu—we interviewed a gentleman by the name of Mr. Kitamura. It was a difficult time for him, we learned, as he told us about a freak encounter with a bear who had attacked and killed his wife. The couple had been gathering medicinal plants in an area where bears were not usually found. Mr. Kitamura was about a hundred feet (30 m) away from his wife when he heard her scream. She was killed in a matter of seconds by a female who apparently mistook the woman for her cub in the bushes. The bear's reaction when she realized her mistake was sudden and extremely violent. However, Mr. Kitamura was adamant that the mother bear should not be caught and put down. His own son had lost his mother, but that was no reason to make the bear cub an orphan too, he felt.

Of course, cubs do sometimes have to fend for themselves if their mother is killed. This was the case for Cannellito, a young cub whose mother, Cannelle—the last remaining female native to the Pyrenees—was killed by a hunter in 2004. He was ten months old when his mother died. Few thought he would live, but my colleague Jean-Jacques Camarra maintained that the environment was safe enough to be conducive to his survival—and he turned out to be correct. The absence of other bears certainly helped Cannellito to avoid many unpleasant encounters. The only bear he might have crossed paths with was his father, a bear of Slovenian origin born in France. Cannellito's case is particularly interesting for the questions it raises. How could he fend for himself at such a young age that he may not have even been weaned yet? How did he prepare for—and survive through—his first winter? Had his mother already made a den before she died, on the first day of November? So many questions we'll never have answered, but they make us wonder at the capacity for adaptation of these great mammals.

In 2013, Cannellito was suspected to have attacked a flock of sheep. In 2017, he is thought to have triggered an automatic camera that captured footage of a big, strong, and healthy male estimated to be around thirteen years old. Unfortunately for him, there are no females left in his territory of the central Pyrenees, at least until more are released. Bears are on the verge of dying out in other parts of this mountain range, and if this happens, the genes of the Pyrenean bears will be gone forever.

By contrast to Cannellito, an unemancipated polar bear cub would, I believe, find it impossible to survive the death of its mother. The rigors of the Arctic environment are simply too harsh. I can't imagine how a ten-month-old polar bear cub would even begin to tackle its first winter completely alone, without a den, in the howling wind and freezing cold. The pack ice is solid in the winter, of course, but it's not easy to hunt seals because daylight is in such short supply. What's more, the only effective technique for seal hunting is to lie in wait by a breathing hole, but it takes a great deal of strength to kill a seal and pull it out onto the ice through the small hole. In fact, it seems that the emancipation of polar bear cubs depends on them reaching a certain size. In April, baby ringed seals, born in a den their mothers have dug in snowdrifts on the pack ice, are the polar bear's prey of choice. Young polar bears must be big and strong enough to smash in the roof of one of these snow caves alone; until then, they'll remain by their mother's side.

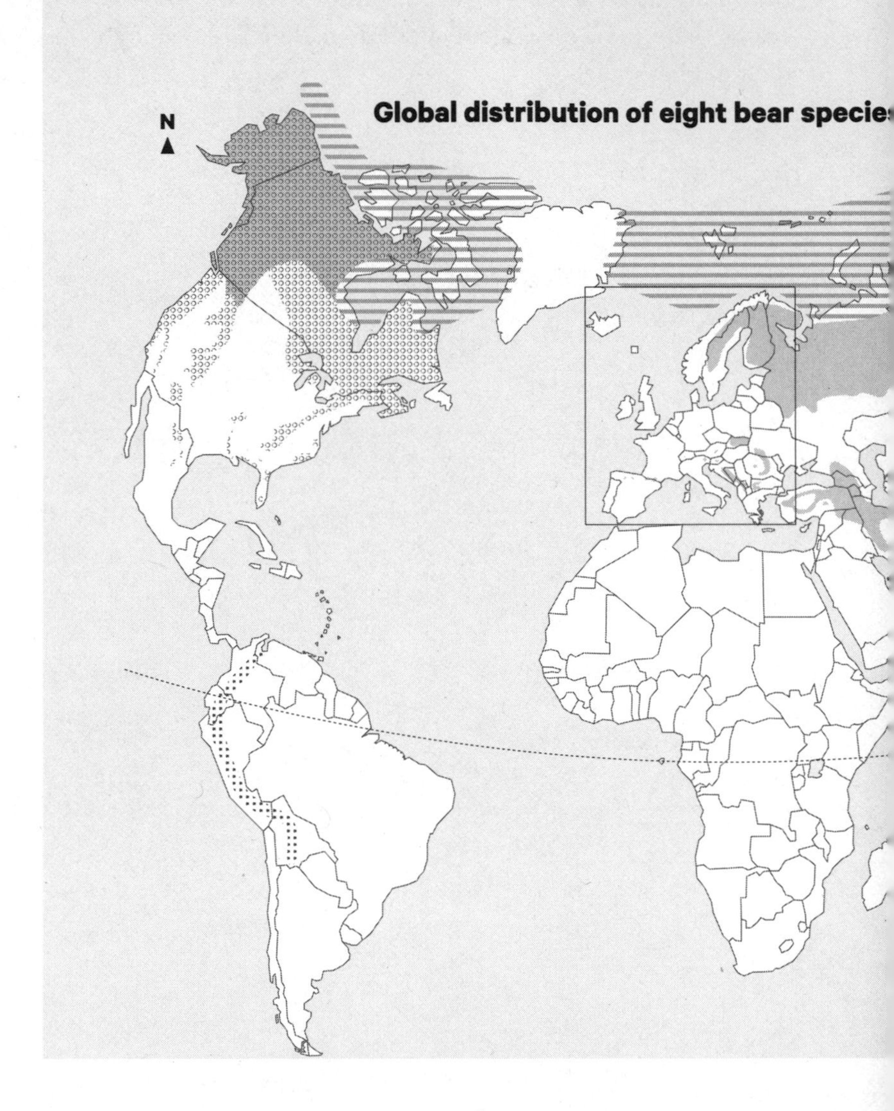
Global distribution of eight bear specie
N

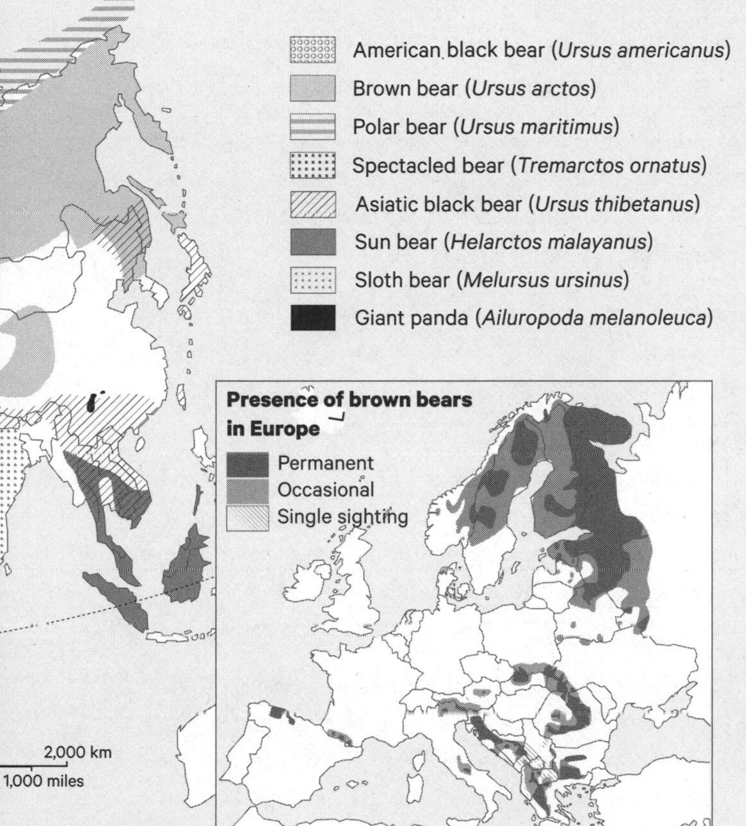

American black bear (Ursus americanus)
Brown bear (Ursus arctos)
Polar bear (Ursus maritimus)
Spectacled bear (Tremarctos ornatus)
Asiatic black bear (Ursus thibetanus)
Sun bear (Helarctos malayanus)
Sloth bear (Melursus ursinus)
Giant panda (Ailuropoda melanoleuca)
Presence of brown bears in Europe
Permanent
Occasional
Single sighting
2,000 km
1,000 miles
urces: ursides.com, reddit.com

Meet the eight members of the Ursidae family

There are currently eight species in the Ursidae family. Details of their habitats, diets, and numbers are as follows:

American black bear (*Ursus americanus*, Pallas, 1780). Some 850,000 to 950,000 of these mostly vegetarian omnivores are spread across the forests and mountain pastures of North America.

Brown bear (*Ursus arctos*, Linnaeus, 1758). These omnivores can be found in boreal and alpine forests across the Northern Hemisphere. Their population is estimated at 200,000 to 220,000.

Polar bear (*Ursus maritimus*, Phipps, 1774). Dispersed around the Arctic from the fifty-fourth parallel north to the North Pole, these pure carnivores prey mainly on seals and small cetaceans. Their numbers are estimated to be in the region of 25,000 (IUCN, March 2017).

Spectacled bear (*Tremarctos ornatus*, F. Cuvier, 1825). This tree dweller eats a diet of mainly fruits and vegetables and can be found at altitude in Venezuela and Peru. With an estimated population of 10,000 to 20,000 individuals, this species is classified as vulnerable and under threat from extensive farming.

Asiatic black bear (*Ursus thibetanus*, G. Cuvier, 1823). These omnivores live in the deciduous and coniferous forests of Asia and can be found at altitudes of around 15,000 feet (4,500 m). Current population estimates are unreliable, but their numbers appear to be in decline.

Sun bear (*Helarctos malayanus*, Raffles, 1822). Native to the tropical forests of Southeast Asia, this insect- and fruit-eating bear is threatened by deforestation and by humans adopting them as pets. Their numbers are unknown, but the species is classified as vulnerable and the population is likely in decline.

Sloth bear (*Melursus ursinus*, Shaw, 1791). The sloth bear is another insect- and fruit-eater that can be found in the forests and prairies of India, Bhutan, and Nepal. No numbers are currently known, but again, the population is likely in decline.

Giant panda (*Ailuropoda melanoleuca*, David, 1869). The giant panda eats an exclusive diet of bamboo, and with a population of only 1,864 individuals (IUCN, November 2017), it is the most endangered of all the bear species. However, its numbers are thought to be rising thanks to drastic protection measures.

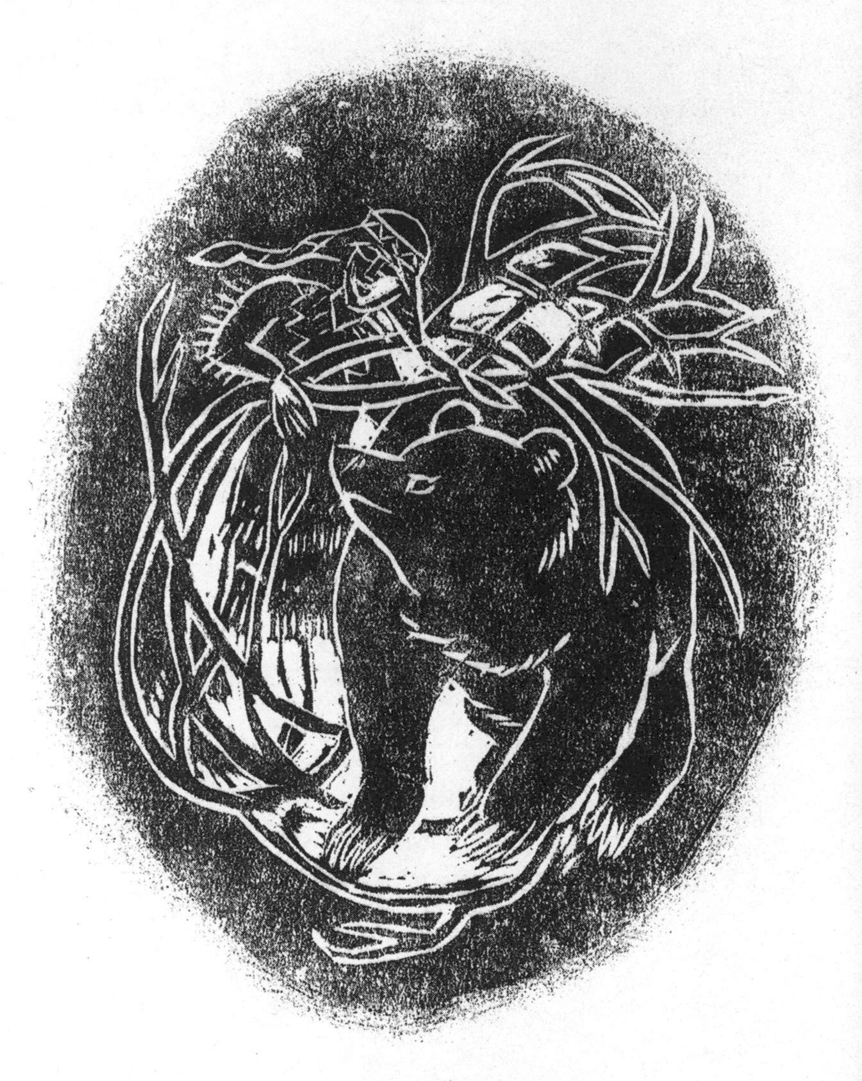

Artwork depicting a woman as a bear's mate

CHAPTER 3

How to Live Like a Bear

A single, invisible bear can transform an entire range of mountains. It casts every peak in a different light, revealing the hidden shadows behind every bush, unveiling the true depths of every forest, reviving the habitats that lie within.

Baptiste Morizot[1]

A BEAR'S LIFE FOLLOWS a natural rhythm according to the availability of food throughout the year, from the moment it emerges from its den, in April or May, until the time comes to hibernate once more, in October, November, or sometimes even as late as December. For adults of breeding age, these six to seven active months are punctuated by mating season, typically from around the middle of May to the middle of June.

Bears live in a variety of habitats. They may inhabit a relatively insular environment amid a forest of foliage, or they may wander the great wide open in the mountains. They are guided in their quest for food by their sense of smell, and are drawn to eat everything from vegetation to animal carcasses and even insects.

Nose to the wind, they roam their territory, foraging their way from one wasps' nest and raspberry bush to another, every course of their alfresco meal invisibly written in scents on the breeze as if they were dishes on the menu of a gourmet restaurant. Bears have such sensitive noses, they'll be drawn to some smells from miles away. For instance, as we saw in the last chapter, the carcass of a whale on the coast of Alaska can attract dozens of polar bears and grizzlies from all around the area. To illustrate just how refined a bear's sense of smell can be, Canadian naturalist Andy Russell likes to quote an old family adage: "When a grizzly sniffs you out, he can tell you the color of your grandmother's wedding dress."[2]

In all my wanderings around the world to observe bears, I've had my fair share of frightening experiences. One time, near the small town of Hyder, which straddles the border between Alaska and British Columbia, I was one of a group of photographers waiting to snap the perfect shot of bears fishing for salmon. *Waiting* was the operative word, because there had been no bears in sight for hours. Getting hungry, one of my fellow observers had the bright idea to cook up some bacon and eggs in the bed of his pickup truck. Before they had finished cooking, along came an excited young grizzly, bounding his way among our tripods and 4×4s. The unsuspecting cook quickly put away the bacon, and everyone retreated to the safety of their vehicles. Well, everyone except me, as I didn't have my own set of wheels, and frustrated by the absence of food, the bear started to follow me instead. Fortunately, and to my great relief, he soon realized I didn't smell anything like the meal he had been deprived of, and he ended up ambling away toward the river.

Bears may sometimes stand up on their hind legs. Contrary to what you might think, it isn't necessarily to get a better vantage point over any long grass or bushes that might be in the way, but rather to better detect the source of a smell or a sound that has drawn their attention. Bears can actually smell and hear far better than they can see. Their daytime eyesight is only about as good as a human's, covering just a short distance. However, like many other vertebrates, bears have what we call a *tapetum lucidum*—a reflective layer at the back of their eye that amplifies light and improves their night vision. Bears may be nearsighted, but they can see far better than us in the dark.

Their hearing is also far better developed than ours. Even the slightest noise can be enough to stop them in their tracks. If they're unable to identify a noise and if they hear it again, they may move away without investigating where it came from. A clap of thunder can certainly give them a fright.

In another memorable experience, I was at a salmon-fishing cabin in Japan, where my colleagues and I were bunking down in the attic. Before I retired to my sleeping bag for the night, I went outside to stargaze for a moment. I had barely set foot out the door when I heard a loud *oof* and a gnashing of teeth, which reminded me that bears tend to be very active in the nighttime. As I turned toward the source of the sound, my headlamp cast its beam on three pairs of bright, shiny eyes belonging to a mother bear and her two cubs. Needless to say, although they didn't seem to show any interest in me, I beat a hasty retreat inside the lodge and only dared venture outside again at daybreak after a fitful night's sleep.

To sum things up, bears use their sense of smell to search for food, their eyesight to take in their surroundings, and their hearing to stay alert.

What do bears eat?

BEARS EAT A varied diet. Because they depend on whatever is available to them in the moment, everything is fresh and in season. It's all organic, too, except for any garbage or junk food they manage to glean from humans. Foodies might say they're true locavores!

In May, when many bears emerge from their dens in the mountains or the boreal forest, their habitat may not yet have cast off its thick winter blanket of snow. Other animals, ones that don't spend the winter in a den, will have had to survive on the meager resources available. It can get extremely cold in these environments, with temperatures frequently well below zero degrees Fahrenheit (-20°C), pushing many animals to their limits.

Every winter, many elk, caribou, deer, and bison succumb to the cold or die of exhaustion, and wolves feast on their remains. When the snow starts to melt, the carcasses of the winter emerge, providing bears with an easy and reliable source of nourishment. Generally, bears will wait a few days after they emerge from their dens to start eating again. They need to eliminate the plug of fecal matter in their intestine in order for their digestive functions to start up again after being dormant all winter long. Breaking wind is a natural part of this process, and unsurprisingly—as we humans can't resist personifying bears and equating their behavior with our own—this has stimulated many an imagination. In Europe, for instance, an old wives' tale claims that bears free the souls of all the babies born during the winter when they release the gas from their intestines in the spring.

When a bear emerging from hibernation sniffs out an animal carcass, it will often drag it out of the melting snow into the open. If the carcass is still frozen solid, this is no mean feat, though the bear may make it look easy to haul potentially hundreds of pounds of flesh a great distance in spite of its six months of fasting and inactivity.

Other times, bears may emerge from hibernation and find frozen berries from the previous year to feed on. They won't hesitate to pick a bush clean of berries, because any food is good to eat. They'll also steal anything they find. For instance, if they come across a stash of tubers that voles have stockpiled for the winter, they won't think twice about helping themselves. I once saw a polar bear do precisely this when it raided the reserves of a group of arctic foxes.

As soon as the snow disappears, the vegetation can flourish. This is when females choose to emerge from their dens with their cubs. The time is ripe to teach them the ways of the world. The flora bursts to life with a sheer variety that is astounding. Brown bears in the Pyrenees get to feast on more than thirty species of plants, while their North American cousins may have more than two hundred to feed on. There are far too many to provide an exhaustive list, but here are just a few:

- Umbellifers (plants from the carrot, celery, and parsley family), including hogweed, pignut, angelica, and Queen Anne's lace
- Buds, flowers, stalks, roots, and fruits of plants including fireweed, rhododendron, crocus, and hawksbeard
- Berries of all kinds, including raspberries, blackberries, blueberries, cranberries, and mountain ash berries
- Most grasses and some common fungi

Bears also love plums, apples, oats, corn, and beets, and plenty more crops that farmers work hard to cultivate. In many agricultural areas, bears are notorious for trampling through orchards and fields and helping themselves to whatever food they can find, just like humans working their way through the aisles of the grocery store with a shopping list.

As the Russian biologist Valentin Pajetnov wrote after observing two bear cubs:

> They each grabbed a big handful of stems, opened their jaws wide, and stuffed the whole lot into their mouths, then, with one swift sideways movement, stripped the stems bare with their teeth.[3]

Bears like to sit down in a field or under a tree and stuff themselves without expending any unnecessary energy, except perhaps when the landowner spots them and tries to shoo them away. They do show some restraint, though. While their sources of food might seem like an all-you-can-eat buffet in the great outdoors, bears will only tend to eat two or three varieties of food at the same time, choosing the plants with the highest protein content.

Bears like to turn over stones and rocks, because on the underside they often find plenty of snails, slugs, ants, worms, and other yummy treats to feast on. They aren't picky when it comes to creepy-crawlies—or anything else that ends up in their stomach, for that matter. Even larger rocks and boulders, which the strongest of humans could never lift, can be moved aside by a bear of average size as easily as a human would move a chunk of gravel.

Along the northern Pacific coast, bears turn over rocks on beaches at low tide in search of crabs and mussels. One group of grizzlies in southern Alaska have the quirky habit of digging for clams on the mudflats at low tide and gorging themselves on quantities too large to count.[4] When the tide is in, and before the salmon are running, these bears eat a mainly vegetarian diet.

One year in July, my colleagues and I were in the Great Bear Rainforest of coastal British Columbia, navigating the narrow channels between a string of islands at the end of a remote fjord. The rain was streaming down, and occasionally we would see the head of a grizzly emerge above the tall, wet grass along the shore. While they were waiting for the salmon to come, the bears here were feasting on the vegetation, bunching it together with their claws, shoveling it between their jaws, and loudly chewing great mouthfuls of the stuff. Unfortunately, bears' digestive systems don't do a very good job of extracting all the nutrients from plants, and their scats are often full of whole blades of grass. Bears are able to digest only about 40 percent of the herbaceous plants they ingest, compared to nearly 80 percent of roots and berries. Their large consumption of vegetable matter might have purposes other than nutrition, then. For instance, farther down the estuary we were navigating, the bears had dug up a plant known as skunk cabbage (*Symplocarpus foetidus*). Because this plant's tubers are high in fiber, they have a laxative effect, which helps bears to eliminate the fecal plug in their intestine left over from hibernation.

Salmon are also a staple of the diet of bears in the northern Pacific—everywhere from British Columbia to Alaska, the Kamchatka Peninsula, and the island of Hokkaido. Seven species of Pacific salmon can be found in this vast area, all of them part of the genus *Oncorhynchus,* whose name is derived from the Greek, meaning "hooked jaw."[5] Two of these, masu and amago salmon, live only in Asia. The five commonly found in North America have spawning seasons that overlap throughout the summer, from late July to late September:

- Chinook, also known as king salmon, is the largest and rarest, with an average weight of twenty to thirty pounds (10 to 15 kg).

- Coho, or silver salmon, is smaller, weighing in at around eight or nine pounds (4 kg), and develops a markedly hooked jaw during spawning season.

- Sockeye, also called red salmon, is the most distinctive of all, highly prized by humans for its bright-red color and widely revered as the "fish of all fish."
- Chum, or keta salmon, is also known as dog salmon, a name it owes to the First Nations Peoples of the West Coast, who deem it less flavorful than other species and have traditionally fed it to their dogs.
- Pink salmon, the last species on our list, is also known as humpback salmon because the males develop a distinctive hump on their backs as they make their way to their freshwater spawning grounds.

As a Frenchman who has traveled the world and seen many places where bears still thrive, I can imagine what it must have been like many years ago when the rivers of my country would have been gorged with Atlantic salmon (*Salmo salar*) during spawning season, and families of bears would have lined the banks to fish, when they still roamed free across the country.

Bears can only fish for salmon during spawning season, when the fish swim their way upstream to their spawning grounds. They're ill-equipped for fishing in the ocean, although on occasion I've seen some impatient bears dive into the waves in search of salmon, albeit to no avail.

Adult salmon are fatty fish, so they have plenty of reserves to last the long journey upstream, as they don't eat while they're "running." Bears fortunate enough to have access to fresh salmon tend to be not only far bigger, heavier, and heftier than their peers but also more sociable. At their fishing grounds, they inevitably cross paths and rub shoulders, not to mention boss each other around and sometimes, very wisely, stay out of each other's way. Many of the big, surly males commandeer the best sites, where it's easiest to catch fish, for themselves, and won't tolerate any other bear's presence. Others are more sociable and seem happier to share the same fishing site with their peers. Males and females cross paths and often simply ignore one another. Bears are notoriously solitary animals, but when the salmon are running, they gather by the river like summer vacationers on the beach. They romp their way through the shallows or stand in wait by a waterfall to catch their prey.

Bears can easily catch up to ninety pounds (40 kg) of salmon in a single day while barely moving a muscle. They like to arrive at their fishing site at daybreak, when the bright light and turbidity in the water makes for the most effective fishing.

It's a magnificent sight to witness a bear splashing its way through the water, chasing hordes of bright-red fish.[6] When they're fishing, bears get so absorbed in what they're doing, they seem oblivious to human presence. They like to herd salmon into little pools by the side of the river and dive right into the pile. It's fascinating to see such lively activity, and tempting to think that bears might enjoy pursuing the salmon and making them jump as much as they enjoy eating them. Salmon roe is their favorite delicacy of all. Some bears like to catch a salmon, eat the head first, and then slice open its belly. If any eggs spill out, they keep on munching, and if not, they discard the rest of the fish.

Bear cubs are introduced to fishing at a very early age. They learn to catch salmon that are already injured, and they like to pick at dead and decomposing fish, delighting in the worms that have started to eat away at the carcasses. Gulls and other winged scavengers then pick at their share of the leftovers, and sometimes a bald eagle (in the Pacific Northwest) or a Steller's sea eagle (on the Kamchatka Peninsula) might swoop in for a piece of the action.

While poaching and hunting remain direct threats to bears, their survival is also threatened indirectly by overfishing of their favorite food. When adult salmon are preparing to swim upstream to their spawning grounds, they gather in dense concentrations in shallow ocean waters, which makes them easy targets for legitimate fishing operations and poachers alike. These fish are particularly sought-after for their eggs, the delicacy known as red caviar.

It's a common misconception that salmon are an infinite resource, but that isn't the case. If spawning salmon are overfished to great excess, there will be no regeneration of their stocks. Fish farms have been built to raise salmon fry in order to repopulate certain rivers in Alaska and on the Kamchatka Peninsula, meaning that the bears in these areas are now entirely dependent on human activity for this aspect of their diet.

Three species of bears are known to fish, and it's interesting to compare their respective aptitudes and techniques. The black bears I've observed in southern Alaska tend to adopt the sentry method of fishing, standing in one promising spot and waiting for a fish to swim by or leap within their reach, then snatching it in their powerful jaws and retreating to eat it in peace. This is a more passive method than the technique of choice of their brown cousins. Grizzlies are generally very active fishers, as I've described above.

Polar bears may also fish in rivers, though they're rarely observed doing so. I was once fortunate to film polar bears fishing in the Torngat Mountains National Park, on the northern tip of Labrador, in Eastern Canada. Here there is no salmon run like on the West Coast; rather, it's arctic char (*Salvelinus alpinus*) that swim upstream, in select rivers only. The estuaries where these polar bears tend to fish generally flow out into the North Atlantic. The bears wait until the greatest concentration of fish are swimming up the river before bounding through the shallow water—only a few feet deep—to catch them. Polar bears are not as well-equipped for the task as the great grizzly bears in the Pacific Northwest, since their shorter, curved claws are less effective for fishing. Because arctic char tend to be livelier fish than salmon, too, the bears must grab them firmly in their jaws to keep them from escaping.

Various sources seem to suggest that climate change is forcing polar bears to resort to fishing in rivers—rather than hunting on the pack ice—but I take a different view. This phenomenon is only very rarely observed in this region, which is devoid of human habitation along a full three hundred miles (500 km) of coastline, because there are only a few days each year when the water levels are high enough for the fish to swim upstream, and there aren't many polar bears here in the summertime. What these unusual observations do illustrate, however, is the similarity between all three species—brown, white, and black bears—and one can easily imagine how the black bears of the Labrador tundra might fish side by side with polar bears.

Let's move to a different continent for another example of not only the diversity of the brown bear's diet but also the opportunistic nature of some of its food choices. On Hokkaido in September, the rivers are gorged with salmon, but the brown bears here still like to eat a balanced diet. I was following a pair of siblings under the forest canopy, a few hundred feet away from the bank of the river. One of the bears, the most daring of the two, quickly scaled a tall tree. With a somewhat disconcerting ease, he climbed all the way up to a branch that was only just strong enough to hold his weight. Brown bears have powerful, crampon-like claws that allow them to scale tree trunks the way mountaineers tackle an icy cliff, and this kind of behavior and agility clearly illustrate their climbing ability.

What had drawn this bear to climb so high, it seemed, were the clusters of wild grapes growing up there. Comfortably straddling his branch, he simply reached up and pulled the vines down toward him with all the ease and expertise of the most agile of monkeys. The problem with bears, however, is that they never let you know what they're about to do. In barely the blink of an eye, this bear slid his way back down from his lofty perch—while I was still standing right at the foot of the tree! I had a lucky escape, but I can see the marks of his teeth on the hood of my telephoto lens to this day.

In the fall, bears choose foods that are rich in carbohydrates and fat to help them gain weight and build up their reserves for the winter. Beehives, wasps' nests, and anthills are all loaded with sweet treats that bears gobble up greedily at this time of the year. If they can't find any bees or wasps, they'll happily munch on acorns, chestnuts, and beechnuts instead.

On the Kamchatka Peninsula, at the beginning of September one year, my colleagues and I were observing a female and her three cubs, who had been eating blueberries for several hours. They were like vacuum cleaners, sucking up berries without lifting their heads from the ground, completely oblivious to our presence, their only concern to pile on as much weight as they could to make it through the winter.

Suddenly, the mother ushered her cubs into a dense grove of Siberian dwarf pine (*Pinus pumila*) trees. The next thing we saw was a tall tree shaking violently. The mother bear was pushing the trunk with all her might to bring it down. Then we heard an almighty crack, and the tree came crashing to the ground. Why would she go to such great effort? you may be wondering. The answer is simple: bears love pine nuts. No sooner had the mother felled the tree than she started chewing on pine cones to extract the soft, oily nuts inside. Her cubs bounded right over to join her, disappearing into the thick branches.

Some bear populations eat highly specialized diets. In August, the bears in Yellowstone National Park climb far above the tree line, sometimes up to ten thousand feet (3,000 m), to reach an unlikely treat: huge swarms of moths called army cutworms (*Euxoa auxiliaris*). The extremely steep and rocky slopes here are a decidedly inhospitable environment, even for bears. But bears are opportunists—as we've seen—and they aren't mistaken about the nutritional value of their prey. They make the effort to climb vertiginous slopes to reach these moths because this species has the highest fat concentration in the animal kingdom, at 72 percent of their body mass. A single bear may eat up to forty thousand of these delicacies per day in its quest to ingest the twenty thousand calories a day it needs.

Some grizzlies in Alaska are aggressive predators and will readily attack elk and caribou (in Denali National Park, for instance), and muskox (in northern Alaska). All brown bears can run very quickly over short distances—up to thirty miles per hour (50 kmh). And they're so powerful that once they've caught their prey, they can easily kill it.

Predation of farm animals and domestic pets accounts for a very marginal portion of bears' diets, even for individuals with ready access to this type of food source. For instance, even though the bear population in the Pyrenees is growing, the number of sheep killed in the region has not increased.

Not always solitary creatures

TO LIVE LIKE a bear means living a solitary life—mostly. The only time bears live as a family unit is while they're raising their cubs, and even then, it's just the mother and her young. The rest of the time, bears are just, well, bears. They tend to avoid each other—unless it's mating season, obviously. Males use the scent glands in their shoulder area to mark trees and claim their territory. They stand up on their hind legs, just like

Baloo in *The Jungle Book*, and scratch and rub up against the trunk to leave as much of their mark as possible. It doesn't matter if another bear has already made its mark on a tree; they're rarely deterred from leaving their own. They're also known to drag their claws down the bark of larch trees, gouging long, deep, parallel lines as a sort of signature.

These claw marks are of great interest to us naturalists. They give us valuable information about how tall a bear is when standing on its hind legs, as well as its strength and the size of its paws. We can also analyze the precious traces of hair it leaves behind in the sap that seeps out from the bark, in order to trace its DNA. As you can see, these markings not only serve as boundary indicators for the bears themselves but also provide important waypoints for researchers.

Females tend to avoid males while they're raising their young, because the males might attack their cubs. This kind of behavior is difficult to explain, but according to one theory, if young bears are killed, their mothers will become receptive again more quickly, meaning that the males will be free to mate again.

Some years ago, when Jean-Jacques Annaud's feature film *The Bear* was in movie theaters, I gave a series of presentations to primary school students. The movie was inspired by James Oliver Curwood's novel *The Grizzly King* and featured only trained and animatronic bears, not bears in the wild. I explained to the children than in the real world, a large male who came across an orphaned cub would never take it under his wing, like in the film; rather, he would make a meal of it. That's the way it is if you're a bear. The children seemed to grasp this quite well, but their teachers would often shed a tear.

I've never directly witnessed a bear killing a cub, though I have been confronted with the aftermath. During a long hike up the Ozernaya River, which opens into Kurile Lake, in the southern part of the Kamchatka Peninsula, I was struggling to keep up with my sprightly young guide, Misha, who was all of twenty-six years old and proud to have arrested three poachers after holding them at gunpoint with his pump-action shotgun. He was used to walking a good thirty miles (50 km) a day as he patrolled the area, carrying all his gear on his back.

I was carrying my own backpack, plus my camera and lenses, with my tripod on my shoulder. We were following the trails that bears had made through the dense and otherwise impenetrable forest, and making good progress, despite treading in fresh bear scats and being whacked in the face with tree branches. The rain was running down our backs like gutters, and my boots were squelching wet with icy water, but Misha, ever the optimist, kept repeating with a smile, "I am dry, I am dry." Still, we were enjoying the hike—until we were stopped abruptly in our tracks by the body of a newborn bear cub lying across the trail. When we approached for a closer look, we saw a very clean, round hole in its skull, which Misha believed had been made by the canine of a male bear. The bear had simply killed the little cub and not made any effort to eat its body. This wasn't the first time I had seen a dead cub. On an expedition to the east coast of Norway's Svalbard archipelago, I observed a young polar bear cub who had met with a similar fate.

Infanticide is always upsetting, but bears are not alone in the practice. More than a hundred species of animals do the same—most notably primates, carnivores, and rodents. The practice has also been observed by a number of human civilizations throughout history.

When mating season comes

DEPENDING ON THE region they inhabit, bears reproduce during the period from mid-May to mid-June. When mating season comes around, they hit the pause button on their solitary lives and set out to find one or more partners. When the females go into heat, the scent of their urine signals their receptiveness to nearby males.

It's not unusual to find several males in the vicinity of a single female. But what happens when two rivals cross paths? If one of the hopefuls is significantly smaller, he will typically be forced to back off following a battle of gesticulating and grunting. If both potential suitors are around the same size, however, they'll duke it out. Fights between males during mating season may not last long, but they tend to be very violent and can even prove fatal. The loser must stand down and beat a hasty retreat, taking out his frustration on any trees in his path. Meanwhile, the winner is free to mate with his prize.

On numerous occasions during my travels in Finland, I've observed encounters, clashes, and couplings like these.[7] The first thing that strikes you as an observer, even though you know these things, is how substantial the difference in size is between males and females when you see them side by side, and especially when one mounts the other.

There can be a mating dance of sorts as the male tries to grab the female, who slips free from his grasp but is still waiting for him. The male shows his excitement by rubbing his thighs up against nearby bushes and trees to mark them with his scent, standing on his hind legs and scratching his shoulders vigorously against the trunks. The female steers clear at first, then draws closer to him, only to bound away at the last second. It's a synchronized show, a courtship ritual, that leads to a brief coupling, which will tend to be repeated several times in the days that follow.

The reproductive process will be more effective if a female encounters several males, since the first suitor she meets is unlikely to be the father of her cubs. A single female may mate with one, two, or even three different males. When a mother bear has twins, there is a 15 percent chance that each cub will have been fathered by a different bear, and the chance of triplets having multiple fathers rises to 50 percent. I've seen one mother whose four cubs looked like two sets of twins, so different were one pair in size and coloring from the other.

Throughout the process, it's certainly the female who calls the shots. She's the one who decides which male or males she's going to mate with. And so she should: after all, she alone will have to raise her cubs, teach them the ways of the world, and show them how to live like adult bears.

When these few short days of excitement are over, the male and female go their separate ways and return to their primary vocation: to fill their bellies with food.

Always eager to judge, the Christian religion historically viewed aspects of the bear's natural behavior as transgressions of its divine law, labeling the creature guilty of five of its seven deadly sins: lust, wrath, sloth, gluttony, and envy. There's a logical explanation for the bear's behaviors, of course, and none of them should be thought of as deviant.

Lust: Religion once suggested female bears were guilty of this sin by mating with potentially more than one male, but obviously there's nothing "immoral" in this behavior; it's simply a necessity in order to ensure the success of the reproductive process during mating season. Some moralists have even suggested that bear cubs are born when they're so tiny because the mother is keen to get rid of them as quickly as possible so she can mate again. As we saw earlier, there's an entirely different explanation for this.

Wrath: Bears who are taken by surprise—especially if they're sleeping—are rarely good-humored. I can certainly appreciate that!

Sloth: For bears, there's nothing lazy about resting when there's food in abundance. They don't stockpile it; they simply eat until they're full and come back for more when they're ready.

Gluttony: Based on what I've observed in Slovenia, I have to say that bears do go crazy for honey or any other sweet or alcoholic substance. There's certainly no shortage of anecdotes about bears getting sleepy and sluggish after gorging themselves on apples or plums, which have then fermented in their stomachs! Bears naturally seek out foods that are high in calories and easy to access, and they're no strangers to inebriation, but they're not gluttons. For sure, they might have a bit of a sweet tooth, but don't we all?

Envy: It's true that bears can have a tendency to set their minds on something. When they catch a whiff of someone else's tasty food, they're not easily deterred.

In terms of the power dynamic among bears, there does seem to be something of a discord between real life and human perception. In real life, it may be the females who are in charge, deciding who they mate with and carrying all the responsibility for raising their cubs, but in the stuff of legend, at least among the Peoples of the Northern Hemisphere, it's the males who are revered as symbols of strength and virility.

While many stories of old have painted the male bear as a family man—a father, a grandfather, a cousin—others have conveyed darker, more sexual undertones. Some legends have recounted how a human female, often a young woman, would be abducted and seduced by a bear, give birth to a hybrid bear-child, and never return to live among humans.

According to one legend of the Evenki People in Siberia, two sisters were caught in a storm as they went out to tend to their herd of reindeer. During the night, as they followed the reindeer tracks, the younger sister disappeared. Unable to find her, the older sister looked and looked, but she fell into a bear's den. There she spent the winter, alongside the bear, sucking the pads of its paws. In the spring, she emerged from the den with the bear, who showed her the way home. She went back to living with her parents, only to disappear again. When her mother went to fetch water, she passed the mouth of a cave and saw her daughter inside with two babies—one covered in fur and the other with normal skin. So no one would mock her daughter, she took the bear cub to look after, and her daughter took the little boy. As the brothers grew older, the young man, whose name was Torgani, wanted to test his strength against his bear-brother's. While they were fighting, Torgani killed the bear with a sharp stone he had fashioned as a weapon in lieu of claws. As he was dying, the young bear shared with the humans the ritual they must observe on the Day of the Bear: hunting, butchering, feasting, and laying to rest. The bear's mother did not partake in the feasting.

Variations on this age-old bear-related myth persist to this day. There are two legends of Indo-European origin in the Western world I'd like to highlight. One of them is very well-known among English speakers everywhere—we'll come back to this shortly—while the other is better known in Southern Europe. According to a traditional folktale in the south of France, the half-human, half-bear Jean de l'Ours (John of the Bear) was the offspring of a human mother and the bear who abducted her. When Jean de l'Ours was big and strong enough, he pushed aside the boulder that was blocking the entrance to the den in which he and his mother were both imprisoned. When they returned to his mother's village, he had great difficulty being accepted by the locals because he was so strong and unaccustomed to living among humans. And so he forged a heavy iron cane and set out on a journey, during which he met companions of similar strength.

It's not hard to see the similarities between this popular folktale and the legend of King Arthur.[8] The twelfth-century French writer Chrétien de Troyes was one of the first to chronicle the story of Arthur, who became a legendary figure of folklore. Arthur was said to have been born on or around February 2 (the Christian feast of Candlemas). His father was a monster of a man who, in a story reminiscent of the later fairy tale *Beauty and the Beast,* wore a mask to take advantage of a young woman. Though there is little, if any, mention of his childhood in the different variations on the legend, Arthur, as we all know, became king when he pulled the famous sword Excalibur from the stone. He then went off in search of other warriors like himself and ended up forming the Knights of the Round Table.

While earlier legends recounted stories of gentle hybrid creatures using their bearlike strength to serve humans, the legend of King Arthur is different in that it reflects the wilder side of the bear, telling the tale of warlords and warriors who react violently and unpredictably. In Celtic mythology, warriors would often kill a boar or a bear in order to gain its strength and ferocity. These half-human, half-beast warriors were known as "berserkers" and were thought to be able to shape-shift and become particularly fearsome opponents. The modern saying "to go berserk"—in other words, to go crazy—seems to be a relic of this.

Bear trainers

IN THE MIDDLE AGES, many bears became circus animals, "tamed" by bear trainers who humiliated these once-revered, celebrated creatures by forcing them to imitate human behavior. Hunted down and demonized by the Christian religion, bears were dethroned—knocked from the top of the animal hierarchy. Once revered, these "fallen kings," to quote historian Michel Pastoureau, became relegated to curiosities.

Hundreds of years ago, groups of bear handlers would travel far and wide from one fair to another, between abbeys and castles, belittling their enslaved bears and putting them on display for entertainment value. This practice was associated with travelers from poor mountainous regions in Eastern European countries, the Balkans, Italy, and the south of France, and is thought to have originated with the Romani People, who came to Europe as early as the eleventh century.

In the Pyrenees, becoming a bear trainer was a common response to a lack of work in the mountain valleys. Bear cubs were "plucked" from their maternal den to be raised as animals for slaughter or trained for display. Trainers would burn their bears' paws, cut their claws, and file down their canines before driving rings through their noses and leading them down the road on chains. These bears were then conditioned to stand up on their hind legs and dance on command to the rhythm of a tambourine or an accordion. They learned to associate different sounds with different behaviors by being forced to stand on hot coals or metal plates, and "danced" in sheer pain to the sound of their trainer's instrument. Just a few practice sessions and the reflex was ingrained. There's nothing natural about that. While some may have viewed the relationship between a "tame" bear and its trainer through rose-colored glasses, there is nothing precious or charming about it. To tame and train a wild animal such as a bear is nothing short of cruelty.

Believe it or not, there are still bear trainers out there today, though I'm happy to say that I've never crossed paths with one. In India, some Asiatic black bears have been trained to "dance," and some trainers in Russia, Bulgaria, Greece, and Turkey are known to have subjected brown bears to the same treatment. Fortunately, the European Commission has now banned the practice, which has resulted in dozens of captive bears being moved to sanctuaries or released back into the wild.

Still, in some places, bear trainers continue to draw crowds at medieval fairs, Christmas markets, and circus attractions. This shameful practice should be a thing of the past and can no longer be justified anywhere as a form of entertainment. It's high time for the powers that be—the civic authorities, show promoters, and circus directors who tolerate this practice—to understand that animal welfare is paramount. The pads of a bear's paws are not designed to walk on asphalt for hours at a time, and a bear is not supposed to eat candy for treats.

I should add that trainers who work with bears in captivity for movies and advertising appearances are perhaps more attentive to their animals' needs and give them more freedom, since the context of their work is very different. Still, it must be said that bears are simply not meant to live in captivity.

How long do bears live?

The typical life expectancy of a brown bear in the wild is around twenty years. By contrast, bears in captivity have been known to live for more than forty years. In the wild, there are at least two critical rites of passage in a bear's life: emerging from the den for the first time, and emancipation. When cubs first follow their mothers out of the den at a very young age, they're confronted with all the dangers that can menace a carefree, clumsy little furball weighing just a few pounds. They can so easily go astray, hurt themselves, fall down a hole, succumb to infection, or even fall prey to another bear. Two years after a mother bear brings her cubs out into the world, it's not unusual for her to have lost one or two of them.

By the time young bears gain their independence, they will generally have spent three winters with their mother. Now they'll have to fend for themselves, find their own food, and, most importantly, prepare for the test that will determine their future: their first winter alone.

After that, adult bears have very little to fear, apart from unfortunate encounters with hunters and poachers, though males may also sustain serious injuries by fighting their rivals during mating season. Adult males can inflict potentially life-threatening injuries on one another when fighting over a mate. For instance, a broken jaw or a canine tooth that's knocked out and becomes infected can prevent a bear from eating and bring about death by starvation.

Other than humans, adult bears don't have any direct predators to fear. Wolves may keep them away from carrion—animal carcasses—but won't attack. Their lives may be threatened, however, by accidents: they might be caught up in an avalanche or rockslide, suffer a bad fall, or get hit by a vehicle. They may also be vulnerable to parasites such as worms, though this is unlikely to prove life-threatening. As they age, bears may suffer from blindness or osteoarthritis.

For several years in Finland, my colleagues and I observed an old bear by the name of Bodari. Every year we returned, we wondered whether we would see him again. As he got older, he seemed to have trouble walking, wheezing as he went. He would sit down at the foot of a tree, and every time he did so, we thought he must be close to dying. But whenever a young male drew near, he would spring to his feet and growl so loudly, the whole forest must have heard. When Bodari eventually died, by our calculations, he must have been around thirty-four years old—an astonishing longevity for a bear in the wild.

CHAPTER 4

The Bear in Its Environment

Every spring, we witness the "resurrection"
of the bears. These indefatigable vagabonds wake
from hibernation to roam unspoiled, still-wintry landscapes.
In a matter of days, they lace the mountaintops with a
string of tracks from the deepest crevice to the highest
peak, casting the whole area in its finest light
under the shining springtime sun.

Jean-Jacques Camarra[1]

MANY A WRITER, historian, voyager, and explorer since the Middle Ages has cast light on bears and their place in the Northern Hemisphere. Bears tended to figure fairly prominently in literature from the fifteenth century through to the early 1900s, reflecting the fact that until then they remained relatively widespread—in spite of the systematic, intensive efforts of humans to wipe them off the map. Today, bears still do roam free, of course, but we continue to reduce the size of their habitats and drive them farther and farther away, to higher altitudes and more remote niches, removed from areas of human habitation. Where bears once roamed from coast to coast, their environment is slowly being whittled away.

It's interesting to see how differently bears have evolved on the North American continent compared to their cousins on the Eurasian continent—a vast landmass stretching across a dozen time zones where bears and humans have always coexisted to some extent. In North America, bear populations were first affected thousands of years ago by waves of human migration across the Bering land bridge—now the Bering Strait. Brown bears from Asia also migrated to the Americas via the same route. Much later, around the middle of the nineteenth century, the mass influx of European immigrants would forever change the face of human-bear interactions and devastate the population.

In North America

LIFE FOR BEARS in North America has dramatically changed since the arrival of European settlers, and their food sources are now very different from what they once were. In the Pleistocene era, brown bears lived on the fringes of glaciers amid tundra-like vegetation and alongside other subpolar and polar species. As the glaciers retreated, large herbivores, gophers, and ground squirrels invaded the Great Plains. Millions of bison, pronghorn, and elk once called this vast territory their home, but they were almost completely wiped out after European colonization. As the Great Plains were transformed by agriculture and human-driven development, these once-plentiful food sources were decimated, to devastating effect on the bear population.

When Meriwether Lewis and William Clark set out on their 1804–06 Corps of Discovery expedition to explore the western portion of the United States, the population of brown—grizzly—bears was estimated at around fifty thousand individuals. As European settlers moved in, bears were gradually relegated farther and farther west. Then, as cattle ranchers took over the country, they embarked on a campaign of mass destruction against the bear, paying professional hunters by the hide to rid their land of this predator.

Forestry and mining operations, road building, and habitat fragmentation did yet more damage to bear populations. The decline and ultimate eradication of the grizzly bear from the eastern part of the United States coincided precisely with the trailblazing of the pioneers. Today, grizzlies occupy barely 2 percent of the territory they once roamed, and their population in the contiguous United States has been reduced to just 1,200 to 1,400 bears.

In his marvelous book *A Sand County Almanac,* the American author and conservationist Aldo Leopold puts his finger on the bleak outlook for grizzly bears and the wilds of nature:

> There seems to be a tacit assumption that if grizzlies survive in Canada and Alaska, that is good enough. It is not good enough for me. The Alaskan bears are a distinct species. Relegating grizzlies to Alaska is about like relegating happiness to heaven; one may never get there.[2]

In Canada, the grizzly bear population has always been confined to the west and north of the country, though black bears have thrived east of the Rocky Mountains all the way to the Atlantic coast. The presence of bears in Western Canada has been conducive to the development of a forest cover unlike any other in the world—the Pacific temperate rainforests ecoregion of North America. Lashed with the heavy rains that sweep in on the North Pacific current—which flows from Japan toward British Columbia before moving up the coast toward southern Alaska and the Bering Strait—the trees in this vast forest grow to immense proportions.

It's a relatively recent landscape. The glaciers here began to retreat some ten thousand years ago, and the environment has remained stable for the last five thousand years or so. The glaciers and melting snow have shaped the landscape here, carving and eroding valleys and creating rivers and torrents that paved the way for salmon to recolonize this coast as the glaciers retreated.

As early as the Pleistocene era, salmon were thriving in the watersheds of the Pacific coast, and their presence has evolved with the ebb and flow of the glaciation and volcanic eruptions that have changed the face of the region. Salmon are anadromous fish, meaning that they swim upstream from the ocean to spawn in fresh water each year. This annual salmon run provides a valuable influx of nutrients from the coast toward the interior of the continent. As the salmon make their way upstream, great numbers of bears are drawn to these rivers to feast on the plentiful source of food. Analysis of bear bones found over six hundred miles (1,000 km) from the coast has shown that 90 percent of their carbon and nitrogen content came from the ocean, and it's all thanks to the salmon.

Studies have shown that during fishing season, a bear will carry on average 3,500 pounds (1,600 kg) of organic material from the river mouth into the forest, over five hundred yards (500 m) away.[3] Within this radius, the trees grow up to 60 percent taller than in the rest of the forest.

Scientists have used isotopic analysis of old-growth trees to determine their nitrogen-14 and nitrogen-15 content. The nitrogen-14 isotope is found in the air, while nitrogen-15 is found in the oceans. Twenty-five percent of the nitrogen molecules found in these ancient trees are found to have come from marine resources transported by bears. By examining the trees, it's even possible to identify periods of time during which human presence dissuaded bears from coming to their fishing grounds on Haida Gwaii (formerly known as the Queen Charlotte Islands). As the Indigenous Haida population dwindled in the late nineteenth century following epidemics of smallpox, the bears returned, promoting the growth of taller trees.

According to 2013 estimates, some 2,300 adult and adolescent bears on Kodiak Island in Southern Alaska consume over four thousand tons of salmon every year—that's about 6 percent of the commercial catch. It's not hard to see how quickly and drastically excessive commercial fishing could directly impact the bear population. The next casualty in line would be the vegetation in this region, where the trees provide a valuable wood resource. An overexploitation of the resources in the Pacific Ocean might therefore jeopardize the forestry industry. However, there are still areas of this forest that have never been exploited as a resource by humans. These areas of primary-growth forest are an increasing rarity in the world, and they deserve our full attention, because they are home to a unique biological diversity that can never be rebuilt. Traveling to observe bears in areas of such rare beauty is one of my greatest privileges as a naturalist.

In July 2000, some friends and I were on the West Coast of Canada, navigating our way by Zodiac through the maze of channels and fjords of all sizes that line the shore near Bella Bella, British Columbia. As always, it was raining heavily. As we nosed our rigid-hulled inflatable boat into a new pass, it was as if we were pushing open the door to a cathedral. This place, the whole landscape, commanded our utmost attention and silence. Enormous dead branches protruded from the mudflats, where shorebirds hurriedly dashed along. A frightened gray heron took flight with a splash of dark water. It was a bleak setting, but there was something peaceful about it. Truth be told, we were hoping to catch a glimpse of the rare white-blond spirit bear, or Kermode bear (page 39). There is an air of mystery about this unique creature. As one story from the Tsimshian First Nation goes, when the glaciers receded, Raven the Creator took flight and turned everything green in the lush rainforest of the Pacific coast. Landing on an island inhabited by black bears, Raven decided to make the fur of one in every ten bears that crossed his path white, to serve as an eternal reminder of the beginning of our times.

This was certainly a wild, unspoiled, and otherworldly place, but we didn't see any bears. As we found our way out his little labyrinth through another channel, we couldn't believe our eyes when we rounded a bend and happened upon a huge steel barge loaded with logging equipment. The operator was waiting for the go-ahead from his foreman to start felling, grinding, and ripping these trees to shreds. It was a fearsome machine that would devour anything in its path. This beautiful, natural Sistine Chapel of a forest and its old-growth trees were standing when the great Michelangelo painted his biblical frescoes, and now they were going to disappear from the face of the planet with the simple stroke of a saw blade, all in the name of cold, hard cash.

Fortunately, since 2016, all sixteen million acres (6.4 million hectares) of the Great Bear Rainforest—an area roughly the size of Ireland—have enjoyed protected status, and only 15 percent of the forest can now be harvested and managed, in a sustainable way. For once, the local First Nations People and ecologists succeeded in winning this fight before it was too late, and the spirit bear was the battle horse they rode to victory. Thanks to this special bear and its rare presence, they were able to protect some of the last remaining old-growth trees for many years to come, if not forever. When a bear has the power to protect a region like this, the whole ecosystem reaps the benefits, as do the people who live in its midst and those fortunate enough to pay a visit.

In Eurasia

BEARS ARE FREQUENTLY present in the lore and literature of the Eurasian continent. Many hunting stories and encounters with the Indigenous Peoples of Siberia have yielded tales of extreme bravado and frankly shocking numbers, with some recounting the killing of dozens of bears.

In his book *Across the Ussuri Kray: Travels in the Sikhote-Alin Mountains,* Vladimir Arsenyev wrote:

> In European Russia, a solo bear hunt was considered a heroic feat, but here one-on-one encounters with bears were a matter of routine, even for the youth. Nekrasov [the poet] extolled a peasant who killed forty bears, yet here were the Pyatyshkin and Myakishev brothers, each of whom had killed more than seventy bears apiece on solo hunts. Then there were the Silins and Borovs, who had each killed several tigers and had long ago lost track of their bear count.[4]

Regardless of how many of these trophies were Asian black bears and how many were brown bears, the numbers are simply staggering.

I assume that if you're reading this book, you're into nature and the great outdoors like me and will forgive this short digression. Some colleagues and I were at the Russian Geographical Society in Saint Petersburg, researching material on artists of the polar region. After a while, a petite *babushka* of a woman came over to chat with us, and our conversation soon turned to Arsenyev, the great Russian explorer. Out of her desk drawer, she then pulled an old black-and-white photograph of a group of Russian soldiers. "This is Arsenyev," she explained, "and on his right is his guide, Dersu Uzala. And here, the youngest of them all, is my father, Otchakov. He was fifteen years old when this was taken." Uzala figures prominently in Arsenyev's fascinating travel writings, which inspired the Japanese film director Akira Kurosawa to make the cult motion picture *Dersu Uzala: The Hunter*.

First hunt of the winter, Bear-Hunting Club of Saint Petersburg, Russia (1906)

But let's get back to the topic of bears. We need to shift our frame of reference to understand the presence of bears in this part of the world, since their population density is so considerable. The estimated global population of brown bears is between 200,000 and 220,000, and Russia alone is thought to be home to some 120,000 of these. These numbers must have been considerably greater—potentially as high as two to three million individuals—before the population was hit by major waves of destruction starting from the Neolithic era onward. Of course, these figures are only estimates, as there is no way to check, unless genetic research can one day tell us.

These large omnivores have changed the shape of the plant life in the Eurasian mountains that are now their last refuge, and they've influenced the way people live here too. I've been to some mountainous areas where the soil had been turned over so thoroughly, I wondered what could possibly have been responsible. There are no wild boars at this kind of altitude, so it must have been bears. As bears dig for one of their favorite delicacies, pignut tubers, they unwittingly give nature a helping hand by aerating the alpine meadows, fertilizing the furrows with their scats, and sowing new seeds. Who knew they were such efficient gardeners? Bears also turn over rocks and crush decaying tree stumps in search of grubs and centipedes. In doing all of this, they change the landscape.

Bears are creatures of habit. Because they tend to always travel the same routes, they effectively create pathways as they tread the ground year after year. These great walkers follow the contours of the terrain, crossing mountain passes and traversing valley bottoms to get between feeding, mating, and resting sites. Using these well-worn trails to get from A to B helps them conserve energy, steer clear of humans, and move as stealthily as possible. I can't help but think that the farmers and shepherds who complain about bears today and want to banish them from their land like their ancestors tried to do should remember that the first settlers in the Eurasian mountains likely followed bear and wolf trails to get where they needed to go, too.

As the American author Doug Peacock has recounted regarding his search for grizzlies:

> There were at least three new sets of tracks, of big grizzlies. I slowly worked down to the meadow where the grizzly family had appeared the day before. Across the flat, I could see four well-used trails in the morning dew. These trails were curious: they looked as if someone had run a double tire track, four inches apart, across the meadow, which merged as it entered the woods to the south. From reading sign in moist areas, I could tell these were grizzly bear trails. I explored two nearby meadows, finding a similar set of trails all leading to an area south of the creek. I remembered reading the Russian zoologist Middendorf—after whom the subspecies of grizzly living on the Alaskan coast is named—telling of brown bear trails in the Siberian forests. The bears walked in the same tracks year after year, and the thousands of superimposed paw prints cut narrow, deep trails so uncannily resembling human paths that the remote forests seemed haunted with invisible people.[5]

This story makes me think of a time when my colleagues and I were tracking bears near Magadan in Eastern Siberia, an area where many prisoners of the Soviet regime were sent to work in forced-labor camps. Twenty-one million people were sent to this city under the Gulag system and never went home again. We saw the trails that the prisoners had forged through the forest here as they moved around. One, as wide as a two-lane highway, was peppered with bear tracks all the way across. If humans can follow the trails bears have made, why wouldn't bears follow the paths humans have made? This trail told us so much about the local bear population, it was like an open book. We could clearly see that bear of all sizes, from large males to females and their cubs, had passed through here. I could just imagine crowds of bruins hurrying their way noisily down this busy avenue, leaving their tracks in the thick, black mud and their scent floating on the breeze. Except that bears, being solitary creatures, would prefer to travel alone and in silence, so to see this many tracks here meant that they had been treading this path for a while.

One might say that bears showed humans the way. In all likelihood, back when their numbers were far more significant than they are today, bears played a role in establishing the paths that were followed by the first explorers, farmers, and hunters.

When individual bears from Slovenia were introduced to the French Pyrenees, the scientists in charge of monitoring them soon observed that they followed the same paths as those taken by bears that were native to the region. We're not just talking about paths on the ground, but also olfactory trails left behind by generations of bears that have traveled these ways. Like a map traced by claws, these heritage paths are there to show the way for the bears of the future.

But one might wonder: Are the bears we observe today the same as those of fifty, a hundred, or five hundred years ago?

Surely their behavior must have changed as humans increasingly encroach on their habitat?

Based on our observations from hides in Finland, bears are clearly influenced by the presence of bait. Of course they are, because they're drawn to food. This means that as naturalists, we have to be very cautious in deciphering bears' behavior in terms of what's truly natural. Behavior during mating season—the male's advances and the female's repeated rejections, followed by the coupling itself—can only be observed when the conditions are right. But the artificial introduction of food can give us some insight about their feeding habits and enable us to observe their behavior in certain situations, such as when females and their cubs are in close proximity to males.

It can be difficult to observe bears' feeding habits in the wild, but salmon rivers provide an easy opportunity for viewing their behavior in situations where food is naturally present. However, it's important to keep in mind that many of the salmon-bearing rivers in Alaska and on the Kamchatka Peninsula are now stocked by human hands to some degree. When the salmon are running, bears gather in extreme proximity to one another, and the females even allow their cubs to frolic near certain males. One can't help but wonder whether the presence of human observers at sites like these is reassuring and contributes to their apparent comfort in being around other bears.

Is there still a place in this world where bears have never come into contact with humans, crossed their logging roads, heard a vehicle or chain saw, or smelled engine oil or gasoline? No matter how hard I've tried, I've come to accept that places like these are extremely few and far between, if they even exist at all. Siberia and Alaska have been roamed for centuries by hunters, gold seekers, and even prisoners. For centuries too, Eastern Europe, and indeed Europe in general, was traditionally more rural than it is today. Since time immemorial, people have grazed cattle in the forest, chopped down trees for firewood, and hunted to feed their families, as well as cutting roads and trails. European bears tend to be more active at the end of the day, after sundown, to avoid humans. Conversely, in less densely populated parts of the world, such as Alaska and the Kamchatka Peninsula, they're perfectly comfortable moving around during the daytime.

Let's not forget that in many places, bears have largely been eliminated by people—hunted down, trapped, or poisoned. We tend to think of bears as wild animals, but they're perhaps more habituated to us than we might assume. They've had no choice but to adapt to human civilization in their natural habitat, by retreating farther and farther away, moving to higher altitudes, and learning to avoid cars, roads, farmers' dogs, and the sound of hikers' boots. This raises the question: What is their natural habitat nowadays?

Obviously, these days, it is not uncommon to see bears on logging roads and trails built by humans. These simply make it easier for them to get around. Footage that was captured from a video camera attached to the collar of Tolosa, a five-year-old female bear in Slovenia, clearly shows how bears seek to avoid humans as they move from place to place. For a month, the camera recorded Tolosa's comings and goings, her play, and her food-sourcing habits as she foraged for berries and dug for beetle larvae, among other things. The footage also revealed how this bear waited in the forest for a vehicle to pass before crossing a logging road, and how cautiously she approached a farm building. Bears are intelligent creatures. They're perfectly able to think, learn, and adapt to their surroundings, when given the time to do so.

Bears and dogs

THE FIRST TIME I traveled to Finland, I was on an expedition with the artist and naturalist Éric Alibert, and our guide, Kai Nyholm, was using a specially trained Karelian bear dog to lead us to the bears. Wading through peat bogs, looking for signs that that a bear had passed by, we first saw a demolished and raided wasps' nest, then a tree stump that powerful claws had ripped apart to extract the tasty grubs. Elsewhere, we saw an anthill that had been reduced to a pile of twigs, a few frantic ants still milling around squirting formic acid, clearly angered that a big clumsy bear had obliterated years of hard work with one swipe of its paw and munched on half their colony.

Karelian bear dogs, like other breeds of specialized hunting dogs such as the Finnish Spitz, the Russian Laika, and Japanese Akita, are prized by humans for their keen and sophisticated sense of smell. These are very hardy and resilient dogs. In Japan, I was part of an expedition to observe Asiatic black bears, and we used Karelian bear dogs that had been specially trained by bear biologist Carrie Hunt in Montana. These dogs had been conditioned to be the perfect companions by scaring the bears far enough away from humans for their own protection.

In Churchill, Manitoba, I became acquainted with Mikan, a handsome Laika who could scare away polar bears four times his size. He would howl like crazy to drive them as far away as he could. Bears aren't accustomed to this type of aggressive confrontation, so they generally take off without any fuss and don't come back in a hurry.

I went trekking in the Svalbard archipelago with Zagrey, a Yakutian Laika who was particularly effective at scaring polar bears away. He had been injured by a polar bear on the *Tara* expedition, a research voyage through the ice of the Arctic aboard a schooner, but he was still very aggressive toward bears. Sadly, some years after our journey, a large male bear took him by surprise during an expedition aboard the sailboat *Vagabond*. The bear was so quick, Zagrey didn't even have time to bark. He was found dead beside his kennel in the early hours of the morning.

Just as guides and expedition leaders use hunting dogs to sniff out bears and keep them at bay, sheep farmers use dogs to protect their flocks from predators such as bears and wolves. Not only are dogs such as the Great Pyrenees (Pyrenean mountain dog), Abruzzese sheepdog, and Anatolian (Kangal) Shepherd big and imposing, they're very protective of their masters and their property.

To be effective, dogs like these must be properly trained from a very early age. Many are born right in the sheepfold and grow up with the lambs, one generation after the other specially trained to protect the flock. If a dog of one of these breeds isn't raised in the right environment, it won't be effective against a predator and may even be dangerous to its owner.

A dog fighting with a brown bear in Moscow, in the late nineteenth century

We've seen how bears can be a fearsome presence in the natural world—hence humans using dogs to track them and keep them at bay. We've seen how they're able to adapt to their environment. But what if they themselves were responsible for the changes in their environment? They've traditionally forged paths across the most inhospitable of terrain, but in doing so, have they unwittingly sealed their own fate by essentially paving the way for human settlers to move in and chase them away?

Bears and their peers

WE CAN'T LOOK at a figure as emblematic in the animal kingdom as the bear without also turning our attention to the other carnivorous mammals that share the same environment. Perhaps the bear's two most prominent peers are the gray wolf (*Canis lupus*) and the wolverine (*Gulo gulo*)—similar in name to its well-known fellow predator, yet far more elusive by nature. The gray wolf shares its entire territory with the brown bear, from the plains of the Canadian Arctic all the way to the Mexican borderlands, and from the Cantabrian Mountains of Northern Spain to the Kamchatka Peninsula in the Far East of Russia.

Both species are classified under the genus of carnivores. But everything about their way of life, their annual rhythms, and their dispersion, is different. Wolves move in packs as family units. These exclusively carnivorous predators and scavengers are active year-round. The wolf is the ancestor of all the domestic dogs we know today, and has existed in the midst of humans for some thirty thousand years. The first humans are thought to have fed themselves from carcasses the wolves had left behind, then created the first breeds of dogs by selectively breeding domesticated wolves in order to protect themselves and keep bears at bay.

I've been fortunate to observe wolves on various occasions in different environments, and there's always something special about the moment. The first time was in 1988, during an expedition to Ellesmere Island in the Canadian territory of Nunavut, about five hundred miles (800 km) from the North

Pole. Vegetation is only visible along the shores of rivers and streams, dotted around at the foot of the scree slopes, rocky outcrops, and mounds of arid earth deposited by glaciers and consistently subjected to a harsh freeze-thaw cycle. Wildlife is a rare sight, too, at eighty-two degrees latitude north. That year up there, the sun didn't set at all for nearly four weeks. It's quite something to experience broad daylight twenty-four hours a day. We were there to look for wolves, and after two weeks, we still hadn't seen any. Then one day, we found the carcass of a muskox stuck in the mud by a small pond and made camp there, in the open air, so we could stake it out.

We woke to find a curious arctic fox sniffing at our sleeping bags. Perhaps it was a stroke of luck, or perhaps just a coincidence, but right at that moment, a handsome white wolf appeared like a spirit on the ridge across the valley, standing in stark contrast to the deep blue sky. It was a magical moment, like a waking dream. We couldn't believe our eyes. At last, we had found a wolf! We stayed near the carcass for five days. That wolf—and others, up to four at a time—came to the carcass to strip the meat from its bones. They returned several times, and all in all we were able to observe them for sixteen hours. For me and my companions on that expedition, those sixteen hours will forever remain imprinted on our minds. It was an experience we still talk about now, decades later. At one point, the wolves came right to the foot of our sleeping bags to scope us out, looking us in the eye from just a few feet away.

David Mech's research and Jim Brandenburg's photography of this pack of wolves inspired a whole new generation of naturalists. I've seen a lot of things during my travels, but this encounter with these white wolves is one of my most treasured memories. On Ellesmere, there are no grizzly bears—only the occasional polar bear on the coast. Here the wolves live at the top of the food chain, joined only by the Inuit hunters who live in the south part of this immense and remote island.

There can be interactions between bears and wolves when they share the same territory. According to a comparative study between Yellowstone National Park and Scandinavia, both of these predators target elk and other ungulates, but in different ways.[6] In the spring, brown bears prey most often on young calves, because they're less able to kill an adult animal alone. Meanwhile, wolves will hunt in packs to kill adult elk. The bears will then move in on the carcasses of the animals killed by the wolves, and force the wolves to stand aside while they take their fill. Does this mean that wolves have to hunt more intensively, if bears are essentially swooping in like scavengers on their kills? Not necessarily. Unlike wolves, bears aren't active all year round, so there's no real competition between these two predators—rather an arrangement whereby they share the same resources. This is another example of how bears have an impact on their environment.

Humans have decimated the wolf population just as they have with bears. However, the numbers of these cautious, pragmatic predators are now beginning to increase in some parts of Europe. One of their strengths is their capacity to reproduce and disperse.

As well as sharing territory, wolves and bears share a special place in the human imagination. Both species are united by a common destiny—or fate. Vilified by the Church and hated by farmers, yet revered by ecologists, they crystallize our every human fear and fantasy about nature.

In the Scandinavian countries and on Russia's Kamchatka Peninsula—and very rarely in North America—bears share their natural habitat with wolverines. Their curious Latin name, *Gulo gulo,* which can be translated as "greedy glutton," suggests they are something of a piggish creature, but make no mistake, wolverines are ferocious predators.

Wolverines are the largest members of the Mustelidae family, which includes otters, stoats, weasels, and martens. They typically grow to around eighteen to thirty pounds in size (8 to 15 kg), but can grow to over sixty pounds (30 kg). They look something like a cross between a bear cub and a badger, and the tracks of all three animals look quite similar in the snow.

Wolverines have a fearsome reputation for a good reason. They'll feast greedily on any animals they find caught in a hunter's traps and have been known to attack young foxes and bird hatchlings of all kinds. Their sharp, powerful claws allow them to climb trees with great agility. The Finnish wildlife filmmaker Kari Kemppainen once told me how a wolverine had pillaged an osprey's nest that was perched at the top of an unstable tree trunk more than thirty feet (10 m) off the ground.

In the area around Churchill, Manitoba, wolverines live by the tree line. They're such elusive creatures that they're rarely seen by humans, except when one gets caught in a hunter's trap, as one Cree trapper told me.

In his travel writings from a voyage to the Kamchatka Peninsula in 1859, the French doctor Félix Maynard reflects on wolverines and the elaborate tricks they use to kill reindeer:

> The wolverine has left the streams of the peninsula behind and can now only be found on the shores of the Anadyr and the other great rivers across the continent. This creature is as cunning and calculating as it is ferocious. When it wants to take a reindeer by surprise, it will gather a bunch of moss, which the deer love, and carry it up into a tree. Then, when a deer passes below the tree, the wolverine will drop the moss and the deer will stop to graze on it. The wolverine then seizes its chance and jumps onto the deer's back, scratches its eyes out with its front claws and rips its neck open with its razor-sharp teeth.[7]

As it happens, Maynard essentially borrowed this description word for word from the Russian explorer Stepan Krasheninnikov's account of his travels to the area in the mid-1700s.[8] It's human nature to recount the stories we hear to others. After all, that's how legends and folktales are born.

To the south of where the reindeer herds roam, wolverines, like wolves, were once eradicated by the Sámi People. However, they're now returning to this area, and their population is on the increase once again. There's something cute about the way they amble and shuffle, always on the lookout, ready to stop in their tracks in a split second. Their flat paws act like snowshoes in the winter, and they can travel for miles in search of food.

Unlike bears, wolverines don't hibernate. Instead, they must seek and find food in the depths of winter amid the snow and extreme cold. These scavengers are certainly quick to approach the carcasses that researchers in Finland put out as bait to attract bears, but they know they have to wait their turn. They'll linger in the vicinity while a bear is feeding on an animal carcass and dig in as soon as the coast is clear, then make way again if another bear comes in for a piece of the action. Even if a carcass is stripped bare of meat, wolverines will always gnaw on any bones they can find.

The geography of bears

WHAT DOES THE Canadian city of Kelowna, British Columbia, have in common with Bern, the capital of Switzerland? While geographically distant, these two cities are connected in name by bears. Bern draws its name from the Germanic *Bär,* meaning "bear," of course, while Kelowna means "grizzly bear" in the Syilx/Okanagan First Nation language. In my native France, countless towns and cities owe their names to bears—Ploërmel and Dinard in Brittany, and Berville-sur-Mer and Cambernon in Normandy, for instance—despite bears having been absent from these regions for centuries. The names serve to remind us that bears were present when the Vikings came here. In the United States, there is no shortage of places more obviously named after bears, such as Bear, Delaware; Black Bear, Oklahoma; and Plenty Bears, South Dakota. And there are Beartowns in seven different states, including New York, Colorado, and Kentucky. In Russia, the city of Abakan is thought to owe its name either to *abas,* the word for "bear" in the Khakas language, or *balakan,* which means "skin of the small bear" in Old Georgian. In one way or another, bears have left their mark on all of these places.

Wherever we look, the maps of towns, counties, and countries around the world are peppered with so many names inspired by bears, they look like a child's connect-the-dots picture book. Just think about how many places in the world have names like Bear Island, Bear River, Baie de l'Ours, and Pico del Oso. Humans have named all of these places after bears, be it to celebrate a memorable hunting trophy or mark the location of an unfortunate encounter.

Have you ever been to a ski lodge or, say, a medical center with a name like Bear Landing and wondered whether the name actually comes from the creatures who used to roam that place? Most people wouldn't think for a second that the parking lot they drive into day after day might once have been full of berry bushes and beehives and drawn bears from far and wide to feast every fall before they went into hibernation.

In the summertime, many urban dwellers eager to escape the city flock to campgrounds with catchy names such as Bear's Paw with barely a thought to how they might be treading in the tracks of the bears who once roamed there. Similarly, plenty of winter sports enthusiasts hit the road for the mountains in the winter without realizing that bears likely forged the first winding path up the valley. As we've seen already, to save their energy on the way to their wintering areas, bears have always chosen the most efficient routes—just as hurried skiers strive to do on their way to the slopes.

Sometimes, toponymy is just one part of the story, and there are other things about a place that can tell us about the role of bears in its history. Let's take the small town of Andenne in Belgium, for example, which features a bear on its municipal coat of arms. Every year, during the town carnival, people dressed in bear costumes take to the streets, and when their parade is over, hundreds of little stuffed bears are thrown into the crowd. In the middle of the town, there is a stone fountain with a sculpture of a bear and an inscription attesting to the legend that in the early eighteenth century, a nine-year-old boy single-handedly killed a bear that had been terrorizing the neighborhood.

Diary of a bear observation expedition

Their tracks were crisscrossing back and forth over one another's. Bears, wolves, and wolverines have all been here, and not long ago. Anything might happen. We retreat into our log cabin. Éric gets his sketchbooks and pencils ready, I prepare my camera housings and lenses. The shadows are creeping across the snowy marsh. It's a cold night. It's well below freezing inside our hide, and Éric is snoring like a train. A hot cup of tea or soup would be great right now. Little by little, my eyes get used to the darkness and I can just make out the contours of the landscape. Trees, bushes, dead trunks, boulders. No movement to report yet. There's nothing quite like keeping watch in the dead of night. It's easy to get lulled into calm contemplation, but the slightest rustle, the softest creak jolts you wide awake. The next day, we see the wolves passed very close by in the night. Too bad we missed them. So we zip up our parkas and settle down for another night in the hide. It's just as cold, just as dark, and we keep watch just as keenly. A crow lands on one of the carcasses we've put out to attract predators. Time stretches on and on. We lose all sense of the minutes and hours passing. And before we know it, the sun rises again, casting its cold, purple light on us. There's still no movement out there on the plain. They never came. Not a single wolverine, wolf, or bear. Never mind, we'll come back another time.

I wrote these notes in my diary in 1997, but I could just as easily have written them in 2012, 2014, or 2017, because I returned to the same place on several occasions, at different times of the year—in the late winter, early spring, and fall. I've spent many a day and night in these hides in Finland, and with the right clothing, it can actually be quite comfortable. But it does take some patience to spend fourteen hours sitting in a lawn chair in freezing temperatures waiting for the silhouette of a wolverine to enter your field of vision. Staring intensely at the landscape for hours at a time sends you into a trancelike state. Time stands still, and your thoughts wander. I tried for years to catch a glimpse of a polar bear emerging from its den. Only after a dozen or so trips did the stars finally seem to align. When the bears finally came out, imagine my surprise to see the female was wearing a tracking collar! Let's just say that ruined the moment, not to mention the footage I had been trying to film. As it happened, I would never get another chance to capture an image like that.

There was one very memorable experience when my family joined me on a trip to observe black bears at Anan Creek in Alaska in the mid-1990s. We were dropped off by floatplane at a remote A-frame cabin nestled beneath the trees. We were completely alone on the riverbank as we watched the black bears fishing for salmon by the other shore, the rapid torrent the only thing separating us from them. One night, a big bear ambled over and settled down for a nap right in front of our door. We spoke in loud voices and made as much noise as we could, but the bear wouldn't budge. Eventually, the bear rose from its slumber and wandered away without any hurry. After that, we made sure to be on our guard around the cabin at all times.

I returned to Anan Creek in July 2016 and was greeted with a very different experience. There were rangers everywhere barking safety instructions, and dozens of tour operators putting on a show for eager tourists—who only seemed to want to snap a selfie with the bears in the background and chatter among themselves before rushing back to wherever they came from. It's still a magical place with dozens of black bears to observe, but I won't be going back.

CHAPTER 5

The Bear's Winter of Mystery

All this the bear saw and heard; and who can ever know what strange thoughts passed behind those small, sagacious eyes, or what unfulfilled longings surged through that mighty frame, as he gazed so steadily and so long out upon that, to him, undiscovered country with its far off vistas and its unknown inhabitants. But it was not his home, and he never went there.

Grey Owl[1]

THE BEAR IS the largest of all the hibernating and semi-hibernating animals. Generally, animals that hibernate tend to be much smaller, such as groundhogs, gophers, and dormice. These animals don't rely solely on plummeting temperatures to tell them it's time to go into hibernation; their internal body clock is also synchronized by photoperiod—the amount of light they're exposed to during the day. Individual animals belonging to the same species but living at different latitudes will therefore go into hibernation at different times; the higher the latitude, the earlier they'll hibernate. Weather and climatic conditions do also come into play, however, and can influence the duration of hibernation. This means that global warming is having a direct impact on bears by delaying their annual retreat into their dens for the winter.

When animals go into hibernation, their body temperature drops rapidly to just above freezing, and their heart rate slows dramatically. For example, when a gopher (*Spermophilus* sp.) hibernates, its heart rate can drop from as high as three hundred fifty to as low as three beats per minute. The drop in the heart rate of the garden dormouse (*Eliomys quercinus*) is even more dramatic, plummeting from around five hundred to five beats per minute. When this happens, their platelet count lowers, which helps their blood flow more easily when their hearts are barely beating.

When we look at the brain activity of hibernating animals, we can see that it also appears to slow down, even while continuing to maintain vital functions such as circulation and breathing. In fact, part of the hibernating brain shows symptoms similar to those caused by Alzheimer's disease, with a significant reduction in neural connections—although in this case, the symptoms are reversible. Scientists are studying the mechanisms that come into play in animals to trigger a return to normal neural activity once hibernation is over, in the hopes that the findings may lead to potential breakthroughs in treatment for Alzheimer's patients.

Believe it or not, animals in hibernation wake on a regular basis, sometimes as frequently as every three to fifteen days.

Four species of bears use a den in the winter: the brown bear, the American black bear, the Asiatic black bear, and some polar bears. Polar bears are unique in that only pregnant females will spend the winter in a den, in order to give birth to their cubs. Male polar bears, as well as adolescent females or those with older cubs, don't hibernate, because winter is the prime time for them to capture seals on the vast open hunting grounds of the pack ice.

In the fall, as the days get shorter and the weather begins to turn, signaling that winter is on its way, bears begin to scope out a site for making their den. Brown bears tend to prefer an isolated location, typically at altitude on a wooded slope near their fall feeding sites. It's important to choose the right location, because the altitude and sun exposure of a den can make a big difference when it comes to staying warm and cozy over the winter. Not too steep a slope, to avoid avalanches; not too low-lying, so it won't be damp when the snow starts to melt; and not too close to noisy human activity.

Males and females tend to choose different types of sites for their dens. A solitary male may be content with a rustic shelter affording minimal comfort, such as a natural cave or a hollow beneath a rock. Meanwhile, pregnant females and mother bears with their cubs will tend to seek out a cozier spot, perhaps at the foot of a dead tree or anthill, or they may burrow a den into the ground.

Unfortunately, even if bears take all the right precautions, humans might still stumble too close and disturb their peace and quiet. Heli-skiing is in vogue these days for many winter sports enthusiasts in search of fresh powder, but this practice of shuttling people by helicopter to remote snow-covered slopes can encroach on the habitat of hibernating animals and trigger dangerous avalanches. One such case proved fatal in 1998, when a female grizzly and her two cubs were swept away by an avalanche on Alaska's Kenai Peninsula.[2] This occurrence was documented, but it makes us wonder: How many more bears have disappeared in similar circumstances without a trace?

Grizzly bears often dig their den at the foot of a large tree or tree stump, since the roots can provide a roof structure of sorts. Building a den can take days, or sometimes weeks. After hollowing out a cave-like space, bears will cover the ground with a thick layer of conifer branches or soft moss and grass for comfort. Once their den is ready, they'll return to the important task of fattening themselves up for the winter. At this time of year, snow can easily blanket the ground and a wintry chill can set in, even on sunny days. Still, bears continue on their quest for food. Munching on any roots and berries they can find will help them to quell their insatiable appetite and build up the greatest possible fat reserves. Only sometime after the first significant winter storm will they retire to their dens for the season. Studies in Sweden have shown that the environmental factors that exert the greatest influence on bears entering their dens are snow depth and outdoor temperature.[3] Bears know better than to confront the cold. They know their limits and they know when it's wise to seek shelter.

Bears in some parts of the world, such as the Yukon and the Kamchatka Peninsula, tend to go into hibernation later than many others. Why do these bears wait so long before retreating to their dens? One proposed explanation is the presence in the winter of dead salmon in rivers on the Kamchatka Peninsula that are fed into by water from hot springs. These salmon represent a readily accessible food source for stragglers to take advantage of.

Known as ice bears, these dawdling hibernators are considered unusual and are often feared by locals. Indeed, many of them may be older males who haven't managed to build up sufficient reserves for the winter and are understandably irritable. As the anthropologist Nastassja Martin explains:

> *Cha'attham,* or "the ice bear" in Alaska, and *Chattoum,* or "the bear that doesn't sleep" on the Kamchatka Peninsula (they share the same general attributes) are the names used to refer to brown bears that do not hibernate like other bears in the winter. Because they don't get the same rest as their peers, they know they are weak and exposed to danger. They are ravaged and driven mad by hunger. To make themselves stronger, at the dawn of winter they jump into the river and roll around in swamps and pools that have not yet frozen until their bodies are coated in a thick layer of icy armor that makes them very difficult beasts to kill. Emboldened by their newfound strength and blinded by madness, they sink their teeth into people, which only amplifies their affliction.[4]

When they do retire to their dens, both grizzlies and black bears curl up in a ball and fall into a deep slumber. (If they're disturbed, however, they can wake alarmingly quickly.) Above ground, the relentless snowfall insulates them from the cold, keeps the daylight out, and seals them away from the rest of the world. They won't eat, drink, urinate, or defecate for the next five to seven months. A plug of fecal matter forms in their intestine to block digestive function. All urea is converted to amino acids and is believed to pass through the bloodstream from the bladder to the gut to be recycled by bacteria. Their body temperature drops by as much as ten degrees Fahrenheit (5°C),

In a Biblical scene, children jeering Elisha are mauled by bears. "He turned around, looked at them, and called down a curse on them in the name of the Lord. Then two bears came out of the woods and mauled forty-two of the boys." (2 Kings 2:24)

and their heart rate drops from forty to fifty beats per minute—the norm in the summertime—to just eight or ten beats per minute. Their entire metabolism slows by around 25 percent. Although their oxygen consumption is cut in half, a higher level of hemoglobin in the blood optimizes their functioning. With their vital functions suspended or slowed down, bears can remain motionless for almost an entire month. As we'll see later, during hibernation, bears enter a state that would induce cardiovascular disease, renal dysfunction, and osteoporosis in humans. They grow obese and barely move a muscle for six months at a time, and yet they suffer none of these symptoms.

While bears are driven into hibernation by environmental changes, their emergence at the end of the winter is triggered by physiological changes. The first physiological factor to herald a bear's emergence from hibernation is an increase in body temperature. Next will be a dramatic increase in heart rate, followed by a return to physical activity.

I remember seeing a bear's den in Sweden, just a few steps away from a trail in the forest. It looked so cozy and comfortable, it almost made me want to curl up and take a nap right there and then! There were claw marks on the walls, and the ground was padded with blueberry bush branches. This bear den, like most others, was smaller than you might think: barely three feet high by about four feet deep at its longest point (about 1 by 1.2 m)—the bare minimum of space to conserve the maximum heat.

Polar bears' dens are those I remember most vividly, and the first that springs to mind could well have been the last

I ever saw. In February 1995, I was on a mission to photograph the inside of a den in Churchill, Manitoba. My guide was Moris, a man from the Cree First Nation whose father had been a trapper and polar bear hunter. We were riding our snowmobiles near the tree line to the south of the community, and the deep snow was slowing us down. It was a beautiful, sunny day, and the temperature was a brisk minus twenty degrees Fahrenheit (-30°C). After several hours of searching, some spruce trees that had clearly been damaged by an animal caught Moris's eye. We moved closer and could see the tracks of a mother bear and her cubs in the snow around the entrance to a cavity—a gaping hole in the ground less than twenty inches (50 cm) in diameter. Moris, who was admittedly still a little short on experience as a guide, was absolutely sure that the female had vacated the premises.

As I was preparing to wriggle my way headfirst into the den, something made me hesitate, and I unwittingly knocked a chunk of ice into the tunnel at my feet. This was met with a loud grunt from the packed snow below our feet. Moris and I beat a hasty retreat and jumped right onto our snowmobiles, which we had wisely left running. Observing from a safe distance, we saw a female and her two cubs emerge from the hole. The moral of the story? Check for yourself before you crawl into a bear's den!

Some years later, with Moris as my guide once more—but after seeing the mother bear and her cubs emerge with my own eyes—I finally made my way into a polar bear den. It felt a little muggy inside, though it was impeccably clean and there was no smell whatsoever. There was barely enough room for me, so it was hard to believe that a bear had spent five months and given birth to her young in there. The claw marks on the walls of the icy cavern showed that the bear had scratched at the snow to carpet the floor somewhat. The den that had protected this mother and her cubs from the brutal Arctic winter winds was about six feet (2 m) beneath the surface, but it was bathed in a soft, opalescent light that filtered through the snow. For bear cubs, their mother's den is a little like a marsupial's pouch. It provides a buffer between the womb and the outside world that gives them time to acclimatize to life in the wild.

Perhaps I should have heeded a legend from the Deg Xit'an (Ingalik) People warning against exploring bear dens, as recorded by anthropologist Cornelius Osgood in the 1950s:

> Once a man went out in the fall just before the first snow to hunt for a bear. The weather was cold. He found a bear hole at last, killing the bear and skinning it. Then because it was too cold he crawled into the bear hole which seemed like a nice place to stay overnight. He piled grass over the opening to keep out the air and went to sleep. When he woke up from time to time, he turned over. At last he woke up, but he felt strange. The flesh of his face was drawn tightly over his cheekbones. He listened a moment and could hear flies at the door. It was spring. "Did I sleep all winter?" he asked himself. Then he went out. He found the remnants of his bear meat with flies all over it. He felt very weak and it took him a long time to walk home. The people were

> surprised to see him. They had hunted for him all winter. Someone asked, "Didn't your father tell you not to sleep in a bear hole?" That is why people do not go into bear holes.[5]

In some traditions, especially Germanic, it was inconceivable that a large animal such as a bear could go without eating all winter long, and it was once thought that they sucked their paws during hibernation. The Sámi People traditionally believed that the *Uldas*—little fairylike creatures living beneath the ground—would feed bears during the winter.

Johan Turi recounts one such legend:

> There was once a girl who spent a whole winter in a bear's house. And the Uldas fed that girl too. [...] She slept very well throughout the winter like the bear; [but] the bear was a he-bear and he got the girl with child.[6]

Besides what is clearly the stuff of legend, some surprising things can happen while bears are hibernating for the winter. I can recall the observations of a guard in the Magadan area of Eastern Siberia, who told me how some bears would emerge from their dens in the spring with whole areas of their bodies completely devoid of fur. He suspected that rodents had snuck into the den and pulled out some of the bears' fur to cushion their nests. He showed me the photos he had taken of one of these bears, and I could clearly see that on part of the bear's back, the fur was plucked almost bare. It looked like a moth-eaten carpet. Some lucky mice must have enjoyed a warm, cozy winter, we joked.

As I mentioned earlier, bears can wake during the winter and may even leave their dens to wander around nearby. Cases like this appear to be occurring more frequently with the highly variable climatic conditions we've seen in recent years. Bear tracks have been observed in the depths of the winter in Bulgaria, Estonia, and Italy, for instance. If there's a sudden thaw, a bear might think it's already spring, only to poke its nose outside, see there's still lots of snow on the ground, and return to its den to hit the snooze button for a while longer.

The bear's habit of disappearing in the fall and reappearing in the spring has seemed to fascinate humans since the beginning of time. To cite Aldo Leopold again:

> Each spring, when the warm winds had softened the shadows on the snow, the old grizzly crawled out of his hibernation den in the rock slides and, descending the mountain, bashed in the head of a cow. [...] No one ever saw the old bear, but in the muddy springs about the base of the cliffs you saw his incredible tracks. Seeing them made the most hard-bitten cowboys aware of [the] bear.[7]

As many First Nations and Native American initiation rituals illustrate, the transition into adulthood in various Indigenous cultures across North America is equated to a kind of rebirth akin to the bear's emergence from hibernation. Of all the spiritual traditions in the world, this is one of the most widespread. Ceremonies such as puberty rites, shamanistic initiations, and the initiation of men and women into secret societies have been traditionally practiced by Indigenous

Peoples and tribes across the continent. Rituals vary, just like the myths on which they're founded, but in spite of the differences, there are certain common aspects—prolonged isolation, fasting, and a symbolic death and rebirth—that echo the idea of a bear in hibernation.

Indeed, the first step in many of these initiation rituals involves separating a young person from their village and family. In some cases, they might be taken (or left to find their own way) to a specific place, sometimes a remote location deep in the forest or the desert, and in others, they might simply be sent outside the limits of their settlement. Then, they might be expected to take shelter in a remote hut or cave, or even just beneath a blanket.

As with these ancient puberty rites, ceremonies to initiate adults into secret societies have also traditionally revolved around symbols of death and rebirth akin to the annual hibernation cycle of the bear. According to the beliefs of these societies, their members would be reborn into a different life in another world.

In many villages and communities today, in places including the French Pyrenees, Romania, and Italy, there are still festivals that celebrate the traditional legend of the bear who seduces a young woman or girl. As shocking an idea as this may seem to modern sensibilities, these festivities are simply a celebration of renewal to herald the coming spring, the return of the light, and new life on the way. The men of the village don bear costumes to acknowledge how humans and bears once lived alongside one another, united in a common destiny and a mutual obligation for one to confront the other, but also to join forces in the face of adversity. The return of the bear represents a renewed hope for better days to come.

As historian Michel Pastoureau has written, the Christian religion had to fight tooth and nail to uproot many of these deep-seated pagan rituals associated with the bear. Not only did the Church hunt down and physically destroy the species, it also appropriated the dates of the traditional festivals in order to steamroll the bear out of favor in popular customs.

In the Pyrenees, for instance, the traditional celebration of the bear (*Fête de l'ours*) always began on February 2. It's no coincidence that this day became decreed by Catholics and Orthodox Christians as Candlemas, the holy day commemorating the Presentation of Jesus at the Temple. According to popular superstition, this was also the day that the bear was reputed to emerge from its den temporarily, then to return for another forty days if it saw its shadow—a tradition that lives on in modern times as Groundhog Day. What you might not know is that until the early 1900s, Canadians' animal of choice when it came to predicting the coming of spring was also the bear, which slowly succumbed to the American groundhog in popular folklore.

In Russia, bears are generally reputed to emerge from their dens between March 20 and April 15. Does this sound familiar? Easter, of course, celebrates the Resurrection of Christ—and falls between March 22 and April 25 each year. This Christian holiday was clearly instituted at this time of year to compete with the pagan celebrations of spring and renewal.

The Ainu bear festival

Japan's northernmost island of Hokkaido is an enchanting place. Each winter, its mountainous terrain is cloaked in a thick blanket of snow, and along its north coast the pack ice extends out into the Sea of Okhotsk.

I've traveled to Hokkaido several times—in the summer to observe brown bears, and in the winter to see cranes, swans, and Steller's sea eagles. When you travel to Hokkaido, you can't not be curious about the Ainu, the Indigenous People who lived there long before the Japanese. Only a small fraction of the population now identifies as Ainu (though many Japanese have Ainu ancestry), but the traditional culture lives on in many communities through dance, song, and artisanal crafts. There are representations of bears everywhere for tourists to see, but they're often depicted with a menacing air or a salmon in their mouth—images that share nothing with the Ainu People's original beliefs. This animist people worshiped the bear perhaps more than any group in Siberia.

According to Ainu tradition, every year in the winter, a group of hunters would leave their village to seize a cub from its den and bring it back to the village. It's even said that the cub was nursed by a female slave in the chief's house. When the cub grew too unwieldy, it would be locked in a cage on stilts. Then, at the beginning of the following winter, a grand celebration—a ceremony known as Iomante—was held in the cub's honor before it was sacrificed, or "sent off." This meticulously orchestrated sacrifice was said to capture the bear's spirit so it would look well on the village and ensure a good hunting season.

These festivities lived on in their purest form until the early twentieth century, after which they became rarer as the Japanese government's policy of forced assimilation tore villages and families apart. Eventually the ceremonies remained only as an attraction for important Japanese visitors. In an account of his travels among the Ainu People in 1938, André Leroi-Gourhan wrote that during these festivities, the men would eat bear meat and drink in abundance while the women danced as if in a trance. At the end of the ceremony, the bear's skull would be placed on a ceremonial spear.

I'd like to share a very poetic excerpt from a traditional song, entitled "Song of a bear-cub," that provides some insight into the relationship between the Ainu People and the bear they've raised in preparation for the Iomante ceremony. It has been translated from the Ainu language as follows:

After that
they raised me
with a magnificent upbringing.
During this while, my human father rushed about busily
to brew wine.
After a while, the wine
was now ready,

and crowds of women and crowds of young men
gathered together.
Those who were straining the wine
darted their wicker baskets together this way and that,
and those who were whittling inau
plied their whittling knives together this way and that.
This went on until finally
they said it was time for me to be sent home.
After that my father went about with the young men
and let me play for a while until finally it was time
for me to be given dismissal.
I was worshiped magnificently with trade wine, trade liquor,
and also with Ainu wine,
as well as bundles of dumplings and bundles of inau.
My human father also worshiped Walnut Tree Lady separately with
fine wine and fine inau.
After that, carrying bundles of wine and bundles of inau,
I came home to the place of my divine father
and my divine mother.

Remains of a bear following an Ainu sacrifice, framed by *inau* sticks of birch

CHAPTER 6

The Geopoetic Polar Bear

Not to wax poetic, but this watchdog of the ice is a knight in shining white armor with the strength of twelve men and the finesse of eleven, say the people of the North who revere it... It explores every nook and cranny of the cold, with the sole aim of finding prey to devour.

Pierre Perrault[1]

IT WOULDN'T BE right to go on about the brown bear without looking at the polar bear too. Genetically speaking, the two species are as closely related as cousins. Looking beyond the science, though, while the brown bear is ever-present in the human imagination, the polar bear holds a special, more enigmatic place in our hearts.

Sparsely populating the very top of the world, polar bears roam their territory with an elegant yet casual swagger. Through the fine mist of the Arctic summer, a tiny dot appears on the horizon, gliding with graceful poise across the landscape. There's something in this scenario that always reminds me of the opening scene to the movie *Lawrence of Arabia*, when the camel slowly comes into sight amid the waves of heat rising from the desert. From a distance on the pack ice, a polar bear is barely visible as a small yellowish blur, only slightly less white than its frozen surroundings. The tundra, though it appears as a flat landscape, is in fact a sea of pebbles and dwarf willows, filled with hollows and crests for bears to sneak their way through.

I've lost count of all the time I've spent scouring the horizon for a glimpse of a polar bear approaching from afar. The trick is to look for the one thing that doesn't look like it's always there—in other words, something that isn't a bush or a chunk of ice. The more you observe polar bears, the easier it is to predict their movements. Often, a bear will meander its way through the moraine to find the most sure-footed path. By reading the terrain of their habitat, we can anticipate where a bear might be lurking so we're not taken by surprise if it emerges seemingly from out of nowhere. Robert Hainard once said about the art of sculpting badgers, "You have to think like a badger and move like a badger to capture all the little details." According to the same logic, if you want to catch a glimpse of a polar bear, you have to think like a polar bear.

When your binoculars do eventually zero in on a hefty yellow-white mass on the horizon, the first thing you'll be surprised to realize is how quickly it can move. Yet a polar bear keeps a steady gait, unless it's stressed by a human or a fellow bear that's in too much of a hurry, because its number-one priority is to save energy and avoid overheating. Whether it's following its nose toward the promise of food, or simply in transit between two hunting areas, the polar bear always walks at the same pace—about 2.5 miles (4 km) per hour—regardless of the terrain or substrate underfoot. It's almost as if it's advancing to the beat of a metronome set to just the right speed for it to cover vast distances without wasting a drop of energy.

Polar bears are great walkers, constantly on the move across the pack ice in search of seals. We can learn a lot about the extreme adaptation they've had to undergo in order to survive in the Arctic simply by taking a closer look at the way they move. According to the findings of a team of locomotion researchers at France's National Museum of Natural History, the polar bear's gait plays a role in regulating its body temperature: moving too quickly will cause its body to heat up and use more calories than necessary.

As the bear's silhouette grows slowly larger in the binoculars, its long neck will be the next thing you notice. Smooth and supple, swaying constantly from left to right, and carrying only the smallest of heads, it seems to go on forever. As the bear draws nearer, you'll realize how tall it is, even on all fours. It may seem something like a Greek temple, its legs four majestic white columns supporting the weight of its body. The bear's paws are broad yet sleek, especially those in the front. Its enormous rump will look like a counterweight. The bear need only lean backward and lift its front paws slightly to stand on its hind legs as solidly as if it were rooted in the tundra.

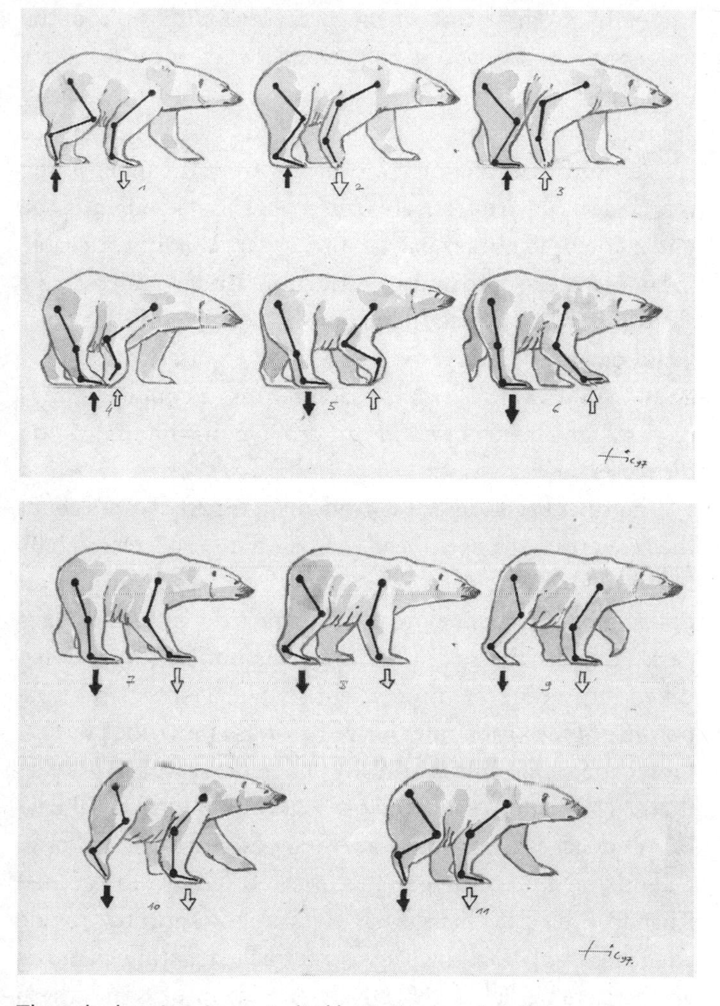

The polar bear's gait, as studied by S. Renous, J.-P. Gasq, and A. Abourachid[2]

You might think that when a polar bear climbs out of the water onto the ice, it would slip and slide like a car on an icy road, but it won't. First of all, the pads of a polar bear's paws are rough and grippy. As the bear plants its paw on the ice, it does so with a rotational movement from the inner to the outer edge, which effectively wrings out the water. Fur on the underside of the bear's paws—this is particularly noticeable in larger males—also improves traction. Under a microscope, we can see that every strand of fur on the underside of a polar bear's paws has fine grooves that act like gutters to channel away the water. This means that after just a couple of steps, its paws are already dry and can provide maximum grip on the ice.

There has been some disagreement in recent decades about the properties of a polar bear's hair. In the 1970s, research by Nils Are Øristland found that polar bear hair acted like an optical fiber by channeling the sun's rays down to the bear's dark skin.[3] By the 1990s, however, his findings were being called into question. Then scientists believed that the sole purpose of the bear's outer guard hair was to provide physical protection, while only the downy undercoat served to provide thermal insulation. In 2016, a new study found that polar bear hair does act as a trap for solar energy, as Øristland had originally claimed. According to this research, led by Mohammed Khattab, even if the hairs themselves are too short to provide adequate insulation, their hollow tubular structure channels and scatters sunlight to effectively provide insulation.[4] These findings have since inspired textile engineers to develop a new type of material that could be used as housing insulation.

As polar bears roam the barren ice, they're guided by their sense of smell and can detect scents from miles away. Polar bears are remarkably capable of sniffing out intruders in their territory—humans included. When they detect a foreign presence, they know it isn't a seal or a walrus, so they're curious. First, they'll sniff the air and squint through their little eyes at whatever has piqued their interest. Next, they'll stand on their hind legs, looking and acting remarkably like an air traffic control tower, to finish sizing up their unexpected visitor.

Once a polar bear has sniffed you out and sized you up, it will use the information it has gathered with its eyes and nose to decide what to do next. Polar bears are inquisitive, and if they've never encountered humans before, they'll casually approach, showing no signs of aggression. Obviously, it's a good idea to let them know they aren't welcome to get too close! If they have come across humans before and have been shouted at, had firecrackers thrown toward them, or perhaps even had rubber bullets fired their way, they'll probably keep their distance. If they're hungry or injured, they'll likely avoid all contact by finding somewhere to observe the intruders and waiting for them to move on.

Whether they're waiting for a seal to poke its head through a breathing hole in the ice or for human visitors to leave, polar bears like to rest their head on their front paw, as if reflecting on their situation with patient contemplation. Keen to be left in peace, they usually just wait for people to pack up their camp and vacate the area. However, depending on how they weigh up the risks of a human confrontation with the potential reward of a tasty meal, they might attack.

One group of young British adventurers in the moraine environment of Norway's Svalbard archipelago found this out for themselves in a tragic incident in 2011, when a starving polar bear attacked their camp. The bear was shot and killed by one of the guides in the group, but not before it was too late. One teenage boy was dragged from his tent and mauled to death, and another who tried to intervene was scarred for life.

During my various encounters with polar bears, sometimes they have come very close—too close for comfort, in fact. One such encounter was in the Torngat Mountains National Park, on the northern tip of Labrador, when I was part of an expedition to film bears fishing for arctic char, which is rare behavior that only one BBC crew has managed to capture on film. In a vast estuary several miles wide, one solitary male bear was wandering by a river empty of fish. Intrigued by our small group, he started ambling toward us, showing no signs of aggression. As the bear drew nearer and nearer—a hundred feet away, fifty feet away, thirty feet away—he still seemed perfectly at ease and relaxed, but we realized we had to do something. We were in a national park, so we had no rifles, only flare guns to scare a bear away. Our guide, Alain, took aim and fired. The bear gave a start, stepped right on the flare that was blazing away on the ground, turned on his heels, and walked off into the water in no great hurry, then came back out and rolled in the fireweed as if nothing had happened. What were the bear's intentions? Was he simply curious, or would he have attacked us? As I often say to people, if you know what a bear is going to do, you know more than the bear does.

In 1922, the French explorer Jean-Baptiste Charcot set a new example of respect for wildlife and the environment when he explained to his crew aboard the *Pourquoi Pas?* that unlike his peers at the time, he would not be hunting the animals he encountered. In his account of their expedition to the Greenland Sea, Charcot wrote:

> The magnificent beast clearly knew that the hunters were so keen to stay on my good side, there was no risk they would shoot at it. With the bellowing of our horn, it eventually got the message and turned its back on our noisy group, sauntering away to the edge of the ice floe, deftly diving into the water and swimming casually over to a more distant floe.[5]

However, only decades later would the values inherent to Charcot's approach trickle down and become standard practice.

In recent years, we seem to have moved far beyond simply raising awareness about these animals. A new era has dawned, with polar bears well and truly in the media spotlight. Have we lost sight of the savage predator behind a veneer of emotion since this majestic bear has become a political and economic weapon for some to manipulate? Everyone has an opinion about polar bears, it seems. People young and old label and nickname these bears as they see fit. *Cute, adorable, graceful, fearsome, ferocious, endangered, facing extinction* are just some of the attributes we assign to these animals. But where does the truth lie? Somewhere in the middle of all that.

Ever-present in the stories of explorers returning from the Far North, polar bears have always embodied the harsh extremes of the Arctic and the dangers that Western adventurers faced in conquering the highest latitudes of our planet. Many drawings and etchings have depicted a man confronting a polar bear standing tall, threatening, and superior in strength on its hind legs, yet the valiant explorer always wins the fight. In reality, a polar bear will never attack while standing on its hind legs, because it will be off balance. If ever a bear did stand off against a sailor armed with a spear, it would only have been because the man goaded it to stand up so that he could jab the spear in more easily. Humans have been manipulating the image of the polar bear for years, and this kind of early depiction serves to illustrate just how long it's been happening.

For the Inuit and many of the Indigenous Peoples of the Siberian Arctic, the polar bear is not a literary metaphor or a communication device. They have lived alongside these animals for thousands of years and have learned to track them, hunt them, eat their flesh, and make a living from selling their hide. Bernard Saladin d'Anglure once wrote that "the polar bear is man's other self." Obviously, the relationship has changed. While the Inuit may not revere polar bears by putting them on a pedestal, these animals are simply part of the way of life in some Nunavik and Nunavut communities. Selling bear hides and guiding foreign trophy hunters can be a significant source of income. There's more to the story, however. Bear hunting is also a way of sharing ancestral knowledge. The bears' habitat on the ice is a harsh environment that is increasingly changing. It takes great determination to venture out there, and to know this place, future generations must learn to "read" the landscape from a very early age. For the

Indigenous Peoples who live on the edge of the modern world, polar bears hold a different kind of economic, social, and cultural significance.

It's nothing new for polar bears to be in the media spotlight: it's been happening in one way or another for hundreds of years. For centuries, the powers in the West have been courting the resources of the Arctic, bringing about a string of conflicts. As early as the ninth century, the Vikings clashed with the Inuit. Later, the Danes, Dutch, and English all fought over the best whaling territories in the Svalbard archipelago. In the seventeenth century, the Kingdom of France and the British Crown battled for control of the lucrative fur trade in Hudson Bay.

Polar bears made regular appearances in the iconography of these conflicts, either in the background or in the form of gifts pledging allegiance. Men would sell polar bears' hides, and many of them would pay with their lives. Trappers and explorers have occupied the vast frozen expanses of Canada, Spitsbergen, and Greenland for centuries, all in the name of profit for their sponsors and governments. Polar bears figured briefly in the very detailed accounts of purveyors of the Hudson's Bay Company in the Canadian Arctic. At the time, the Inuit did not hunt polar bears for trading purposes, instead preferring to keep the resource for themselves. In Greenland, polar bears were regularly hunted until the Second World War by Denmark and Norway, who used hunting to try to establish a colonial presence on the east coast as part of their centuries-long fight to control the island. In the Svalbard archipelago, the Norwegians used to massacre polar bears and keep score of how many they had poisoned, burned to death, or caught in any manner of traps. The trappers' kill counts amounted to thousands of trophies as the carcasses and hides piled up. This senseless carnage bears witness to the dogged determination of humans at the time to eradicate a species that had ruled these lands for thousands of years, in order to make way for other uses.

Just a few decades later, attitudes have changed. Now the rulers of the Arctic have been reduced to promotional tools, marketed in glossy brochures by cruise lines and travel agents to tourists eager to catch a glimpse of a polar bear on the ice. Tourist offices and tour operators are falling over each other to offer the chance of a lifetime to see bears with zero risk and in the best possible conditions. The only problem is, the ice is retreating farther and farther north in the Svalbard archipelago, and the chances of tourists actually seeing bears are becoming slimmer. Once again, the polar bear is becoming a mythical creature, an object of fantasy for tourists who swoon when they see one half a mile away.

This boom in polar bear observation cruises in Northern Europe is partly political: Norway wants to boost economic development in the Svalbard archipelago to counter Russia's coal mine operations in the region. Under the Svalbard Treaty, signed in Paris in 1920, both countries were given the right to engage in commercial activities in the archipelago. Here, the polar bear has become a selling point for one nation's campaign to stake a better geopolitical claim over the land.

Scientists studying the Arctic realized as early as the mid-1960s that something would have to be done to save a population whose numbers had dropped to only around ten thousand individuals. In spite of the Cold War raging and tense relations between the five nations bordering the Arctic Ocean, a number of Russian, American, Canadian, Danish, and Norwegian biologists admirably set their countries' differences aside to join forces and create the Polar Bear Specialist Group in a spirit of research and conservation. Together, they lobbied their respective governments to ratify the Agreement on the Conservation of Polar Bears, signed in 1973. This was the first treaty these countries had ever agreed on. The political boundaries of the Arctic were drawn up relatively recently. Alaska only became part of the United States in 1867. The Svalbard archipelago has been officially controlled by Norway since 1920. Arctic sovereignty in Canada's northern territories came under threat in the 1930s, and remains a priority for the Canadian government to this day.

As Rolf Einar Fife, who has served as Norway's ambassador to France and to the European Union, has explained, the ratification of the polar bear conservation agreement paved the way for the creation of the Arctic Council in the 1990s. Since then, the biologists have been replaced by politicians, and the polar bear that was once a symbol of the Arctic has now become its very embodiment.

Now that the land borders are drawn, territorial claims are extending to the continental shelf. The five coastal nations want to broaden their area of influence into the depths of the Arctic Ocean and, by the same token, claim the right to exploit the economic resources from the sea bed and below the subsurface. Fur trading, whaling, and polar bear hunting are now things of the past. Now oil, gas, offshore fishing, and all the other resources hidden beneath the ice are the bread and butter of international relations in the Arctic.

Since 2006, the polar bear has moved up in the world. Once the vagabond of the North, it's now a global superstar gracing magazine covers, headlining international conferences, and inspiring environmental awareness campaigns around the world. Why 2006 precisely, you might ask? This was the year that former US vice president Al Gore's documentary *An Inconvenient Truth* was released, catapulting the issue of climate change into the media spotlight, with the polar bear playing a starring role. The documentary painted an alarming picture of polar bears drowning in the Beaufort Sea amid increasing greenhouse gas levels and the melting of the Arctic sea ice, all narrated with the confidence and clarity one would expect of a prominent American politician. It's true; climate change is an inconvenient truth. It has also proved to be troublesome as many of Gore's alarming claims failed to materialize in the decade that followed, leading climate skeptics to storm into the breach.

Polar bears have become highly profitable for zoos and other organizations around the world. In 2006, the Berlin Zoo was thrilled to introduce the world to Knut, a polar bear cub born in captivity who was soon rejected by his mother. Raised to maturity by a zookeeper who endearingly bottle-fed the cute white bundle of fluff under the watchful eye of the TV news cameras, Knut became an international celebrity overnight. Business was booming, thanks to Knut. The Berlin Zoo was listed on the stock exchange and visitors were flocking there from all across Europe. The shine began to wear off Knut's story when animal psychologists warned the bear had developed psychopathic tendencies; when Knut died in 2011 right before the eyes of many of his adoring fans, his story came to a tragic end. Still, other zoos, in Denmark, Canada, Russia, and France, have since showcased polar bears of their own. As every new cub is born, a new wave of press releases and marketing campaigns is unleashed. Every accommodation is made to "enrich" the quality of life in captivity for these bears—cold rooms, decor reminiscent of the Arctic, and exercise and play features galore to keep them occupied—but one has to wonder whose lives (or pockets) these efforts are really enriching.

The debate continues as to whether it's appropriate for bears to live in captivity at all. Some prominent biologists have spoken out in favor of the practice, claiming that bears live longer in captivity as they will be cared for and fed and never go hungry. Doesn't that sound like a sweet life? Still, as far as I know, no bear has ever come knocking at the door of a zoo wanting to spend the rest of its life pacing around within four walls, swimming in cloudy water, hearing the incessant noise of screaming children in the background, and waiting to wolf down a pre-portioned meal at the same time every day.

Advocates for keeping polar bears in captivity argue that global warming and conservation of the species are reason enough to continue the practice, even going to far as to suggest genetic mixing as a solution in order to repopulate the Arctic. Judging by their press releases sporting the names of benevolent NGOs, you'd be forgiven for thinking that zoos are trying to have us believe that they offer the only solution to the extinction of the species.

There's a lot of talk about polar bears being endangered, but just how many are there left? According to data gathered over approximately a decade (because polar bears aren't something you can count in a day), the Polar Bear Specialist Group has estimated the population to be between 20,000 and 25,000 individuals. Over the same period of time, various NGOs and zoos spoke out claiming the species was threatened. The numbers may be declining, but it's a big stretch to say that polar bears are on the verge of extinction. In 2017, a new estimate suggested that polar bear numbers were likely between 22,000 and 27,000, with a likelihood of declining significantly over the next fifty years.

The dialogue has to change with the times, because the red flag that worked so well to kick-start fundraising efforts and stir public opinion is no longer effective. Meanwhile, the oil and gas giants seem to be backpedaling on their exploration projects in the Arctic Ocean, but it's too soon to cry victory. Currently the technology to exploit the resources in this part of the world is unreliable, and especially so in a marine environment. What's more, the mere thought of a polar bear dripping with crude oil is enough to keep the boards of directors of oil and gas companies awake at night. Research and exploration are costly, haphazard, and not viewed well in public opinion. It may simply be a matter of time, however, before oil prices skyrocket again and consumers start caring less about the risks. It's amazing how values like environmental protection and animal conservation can fall to the wayside when our purchasing power comes under threat.

They may have become instrumentalized, but polar bears are still magnificent creatures. They're a shining example of an animal that has adapted perfectly to an extreme environment. That said, we shouldn't allow their media stardom to blind us to the plight of other species sharing their habitat that are perhaps even more endangered. Let's not forget that walruses, narwhals, arctic foxes, and caribou all stand to suffer from the effects of rapid climate change.

There's no doubt that the consequences of global warming on the sea ice represent a major threat to the polar bear's survival. However, the risks aren't the same everywhere in the Arctic. While the alarm is certainly sounding for some polar bear populations, such as those in the Svalbard archipelago and around the Beaufort Sea, the prospects are much brighter for those in the Canadian North. That said, the harmful effects of pollution carried north by ocean and air currents, as well as the dangers of disturbance from increased human activity—including tourism, mining, and marine traffic—shouldn't be underestimated.

Today the polar bear stands at a crossroads amid the strategic issues playing out in the Far North. There's no doubt that it has become a geopolitical creature—not only the proverbial canary in the coal mine, but also a flagbearer for goodwill and a catalyst for change. Let's not forget that the polar bear has also become a powerful marketing tool: its image graces the logos of business ventures as varied as diamond mines, zoos, and environmental NGOs. But to describe the polar bear as simply a geopolitical tool wouldn't do this majestic creature justice. Perhaps it would be more accurate to label the polar bear a geopoetic animal. While the poet Kenneth White describes the crow as a bird of many tongues, able to speak to different peoples in their own language,[6] the polar bear speaks to us all in a language all its own.

Casually roaming around at the top of the world as the wind blows the consequences of global events his way, the polar bear seems to be looking down nonchalantly at us from the highest of latitudes. Like a poet, the polar bear is content to do its own thing, indifferent to the tourists passing through, the trucks that rumble by and wake it from slumber, and the helicopters and icebreakers that frighten away its prey. Yes, its future is uncertain, but the polar bear doesn't know that. The polar bear feels none of the human anxiety we feel about the forks that lie in the road ahead and who will join us on our journey.

The polar bear doesn't care in whose service he's being used. He's like Gavroche in Victor Hugo's *Les Misérables,* the young boy who carried the flag over the barricades not knowing why, or for whom. Like a philosopher, the bear poses questions, challenges us, but then continues on his way.

What the polar bear does know is that he has nothing to do with it. His environment, and indeed the world, is changing, but through no fault of his own. And so he reminds us that we humans have only ourselves to blame.

It's clear that the polar bear's habitat is changing rapidly and dramatically. In the last fifty years, the expanse of sea ice in the Arctic has halved in size. In March 2018, its surface area measured around 5.5 million square miles (14.3 million km^2)—

that's the second lowest level recorded since 1979. Nowhere is the retreating ice more apparent than around the Bering Strait. Polar bears have adapted very well as a species to the harsh environment in the Far North, but today they're becoming increasingly vulnerable to rising temperatures in the Arctic and the pollution and interference caused by development and human activity.

The only way to protect the polar bear is to reduce greenhouse gas emissions and slow down climate warming. To do this, we must consume fewer fossil fuels and do so more responsibly. Nothing can prevent this major damage anymore, but we can make a big difference by drastically and sustainably changing the way we consume the resources of our planet.

The birth of a polar bear cub in captivity is not a vision of hope. It is a sign of our failure to ensure the species can live wild and free in its natural habitat.

How polar bears became cash cows for zoos

In the early 2000s, the polar bear population in zoos around the world had been in decline for years. The specimens on display were no longer very interesting—enormous, old, sick bears that did nothing more than wallow in green-tinged water all day—and zoos could no longer simply go and pluck "fresh" bears straight out of their natural environment. Then one day Pandora's box was opened.

When Knut was born on December 5, 2006, at the Berlin Zoo, he was the first polar bear to be born there in thirty years. The event sparked a media frenzy that paved the way for a new public fascination with polar bear cubs born in captivity. This cub was rejected by his mother and only survived by being bottle-fed by a zookeeper.

The day the Berlin Zoo presented the cub to the public, four hundred journalists were there to cover the event, which made waves in the media around the world. Visitors from across Europe flocked to Berlin to catch a glimpse of the polar bear cub, and the zoo's stock price doubled in a week. The Berlin Zoo registered Knut as a trademark, and stuffed toys and other promotional items quickly filled the shelves of its gift shop. This little polar bear cub was featured on the cover of *Vanity Fair* magazine with Leonardo DiCaprio. Knut had a song written about him and a line of popular candies named after him. Meanwhile, his keeper, Thomas Dörflein, died on September 22, 2008, and people barely batted an eye.

A neurological disease led to Knut's sudden death in a seizure on March 19, 2011, with hundreds of horrified visitors looking on. The agonizing experience was captured on video, and a collection was later organized to erect a memorial in Knut's honor. But meanwhile, other zoos were eager to capitalize on his story and reap similar financial rewards.

In 2010, the Columbus Zoo in Ohio opened a new polar bear enclosure to much fanfare after closing its previous facility in 1994, where cubs had been born twice, but with far less media interest. Another media success story was Siku, a cub who was born in captivity on November 22, 2011, at the Scandinavian Wildlife Park in Kolind, Denmark, and taken away from his mother at two days old because, as the press release claimed, "his life was in danger." It's a sign of the times that since the day he was born, Siku has had his own Facebook page, which lists the bear as a public figure and has more than sixty thousand likes. His younger brother and sister, Nanu and Nuno, were later born in the same facility. In April 2018, a zoo in Scotland introduced the world to Hamish, the first polar bear cub born in twenty-five years in the UK.

The number of births in captivity has exploded since 2010. Nine polar bear cubs were born in 2011, eleven in 2012, and another eleven in 2013—and these figures don't include births in China. There's even a calendar of polar bear cubs' birthdays. Every new birth announcement brings a flurry of free publicity to the zoos.

In Missouri, the Saint Louis Zoo has invested US$20 million in a cold-room facility for its polar bears, while in the South of France, Marineland in Antibes has spent €3.5 million (US$4.2 million) on a similar building. These facilities have been so highly lauded in the media that one might almost imagine polar bears in the wild wanting to leave their home on the sea ice to come and live in one of these air-conditioned paradises, if only they knew about them. Of course, it's hard to believe that private enterprises like zoos would spend the money on these investments without calculating their profitability potential. There are also international strategies driving projects like these. One such initiative is the European Endangered Species Programme (EEP), which governs breeding programs for species deemed to be at risk, such as the polar bear.

On December 23, 2014, one of two polar bear cubs at the Columbus Zoo in Ohio was stillborn; the other died just hours after birth. There were two more deaths in 2015, at Ranua Wildlife Park in Lapland, Finland, when the mother ate her newborn cubs. She had given birth to another cub in 2012, who was sent away to a zoo in Austria at an early age. In March 2018, the Cerza Zoo in Normandy, France, took delivery of two young female polar bears born in the Netherlands. A month later, the Singapore Zoo announced the birth of its first polar bear cubs, but also revealed that it had euthanized its adult polar bear at the age of twenty-seven. In 2016, a polar bear born at the Rostock Zoo in Germany, then transferred to Ålborg in Denmark, was euthanized at age fourteen. Wait a second—wasn't one of the main reasons given to justify keeping polar bears in captivity that it would enable them to live longer than in the wild? The problem is, polar bears in captivity tend to very quickly become neurotic and demonstrate stereotypic (repetitive and purposeless) behavior that zoos try to deter by incorporating exercise and play areas into their tanks and enclosures. It doesn't take much to turn the symbol of the savage Arctic into a circus animal.

In 2014, two orphaned cubs, a brother and sister named Blizzard and Star, were found on the ice near Churchill, Manitoba, and transferred hundreds of miles south to the Assiniboine Park Zoo in Winnipeg—somewhat pompously dubbed the International Polar Bear Conservation Centre. Blizzard died in January 2019. Kali, an orphaned polar bear found alone in 2013 in his mother's den in Northern Alaska, has been transferred between zoos in Alaska, Chicago, Saint Louis, and Buffalo, New York, in his short life. Orphans like these are an invaluable gift for zoos, because it's illegal to capture bears in their natural habitat, the only exception being cubs whose lives are in danger. None of these cubs will ever experience life in the wild again.

While in the wild, mothers will only kill their cubs if there is an urgent need for food, in captivity, this type of infanticide seems to be quite common. For this reason, zoos generally separate cubs from their mothers at birth and only reintroduce them at three months of age, when the females are less likely to kill their young. There's a media storm whenever a zoo presents a new cub to the world, to the point where national news organizations have heralded these events as a purely commercial happening, as if each cub were a hot new commodity on the market. Barely a few months after each media announcement, the cubs have grown and are often exchanged or sold and transferred to another zoo or wildlife park looking to enhance its list of attractions.

I should add that zoos and similar facilities may make a valuable contribution to conservation efforts for some endangered species—such as Przewalski's horses or North American bison—by gathering funds or raising these animals in captivity. This is not the case, however, for large carnivores such as bears, which need several months or years to get to know their environment and learn the right hunting techniques. There are some rare exceptions, though. For example, at the Oregon Zoo, researchers have been studying two polar bears in captivity to analyze the absorption of pollutants into their blood. The knowledge they gain will be used to develop a procedure for use in the wild that will help scientists study changes to their diet related to climate warming.

CHAPTER 7

The Bear's Revenge

Seeing a bear was both a dream-like experience for me and a realization of the need for a normal life in a world where I feel connected, a world I cannot believe is gone forever.

Robert Hainard[1]

Skeletal remains uncovered in many caves have shown that humans and bears existed in great proximity to one another as long ago as 500,000 years before the present day. At that time, bears were plentiful and essentially omnivorous, while early humans were hunter-gathers and few and far between. Our ancestors likely lived alongside the cave bear (*Ursus spelaeus*), a close relative of the brown bear we're familiar with today.

We'll never know for sure if there was a spiritual connection between Neanderthals and cave bears. But the bear skull discovered in the Chauvet Cave sitting dramatically on top of a large stone block, and dated at more than 25,000 years old, can't have ended up there by accident. Some authors, including Christian Bernadac, have suggested that the bear was the first god worshiped by early humans, or at the very least served as an inspirational figure, a go-between of sorts between *Homo sapiens* and the forces of nature.

It's impossible to deny the prominence of the bear in traditional cultures across the Northern Hemisphere. Some may be skeptical of the idea of the bear as a god, or even the first god, but the similarities between the various beliefs and rituals of these cultures are striking. Many of them certainly revered the bear as a totem animal, including the Sámi and numerous Siberian and Native American Peoples.

In Western Europe, evidence of bears is widespread among the remains of prehistoric settlements. The bones found scattered in various locations show signs of having been cut, which would suggest that bears were once butchered by humans in a similar way to wild boars and deer. There's evidence to suggest that parts of one bear may have been distributed among different homes, which is not thought to have been the case with other game. It's also worth mentioning that the famous Tyrolean Iceman, Ötzi—a five-thousand-year-old natural mummy found in the Italian Alps—was wearing a bearskin cap and boots with bearskin soles. Until only recently, the Sámi People used to make the soles of their boots out of bearskin to avoid slipping in the snow.

Evidence found at the Grande Rivoire archaeological site, near Grenoble, France, suggests that around six thousand years ago, a bear was tethered there by its jaw. The remains found show clear deformations in the bear's lower jaw, suggesting that the creature may have been domesticated or held captive among humans until around five years of age.

Hunting is an integral part of the historical saga that has unfolded between humans and bears. The earliest evidence of bear hunting by Neanderthals dates back 480,000 years, and I think it's safe to say this has always been the cornerstone of our love-hate relationship, at least here in the Northern Hemisphere. Indeed, our fascination with bears is very likely rooted in our predation of them.

When a dead bear is stripped of its skin, it can look uncannily like a human being. I suspect that even the most seasoned of bear hunters might have found the thought crossing their minds that they had just killed a man. Gruesome as it may sound, the resemblance between the skinned body of a bear and that of a human being must have weighed on the conscience of many a generation of hunters, shaping the idea of duality: the bear as our other, a reflection of ourselves. The pink of the bear's flesh, its four limbs, even its size, conjure an image of a tortured man taken down from the cross. In his book *Le Cantique de l'ours* (Hymn of the bear), author and naturalist Stéphan Carbonnaux cites one young hunter in the French Pyrenees who described the experience of helping to cut up a bear when he was sixteen years old: "When we skinned and butchered the bear, you know, I couldn't help but think it was a man we were cutting up."[2]

As the archeologist, paleontologist, and anthropologist André Leroi-Gourhan explains:

> Just about everywhere, the bear was seen as a man in disguise. That may have been true during the Paleolithic period. [. . .] In all Far-Eastern mythology, and particularly that of Siberia, one can find references to the bear undressing when he goes home, and becoming a man.[3]

The bear hunting carried out by animist peoples was a highly codified practice, and it was remarkably standardized across different cultures. Even though the species of bears being hunted were different—Finnish hunters would go after the Eurasian brown bear, while their Cree counterparts would track the American black bear—their rituals were fundamentally the same. And the Cree and Finns were not the only hunters to share these traditions. The bear-hunting rituals of many of the Indigenous Peoples of North America—specifically all Algonquian language-speaking groups and some groups of Athabascan language speakers—were the same as those of the Indigenous Peoples of Northern Europe and Asia.

We need look no further than how the bear was named to start seeing the similarities. Like the Indigenous Peoples of the sub-Arctic and those of the eastern forests of North America, the traditional Northern European and Asian people avoided using the generic word for *bear* in their language, speaking in synonyms and sobriquets instead. It's as if they didn't want to speak directly of the animal, perhaps so as not to offend the sensibilities of the spirits; Rather, they used names befitting the spiritual intermediary with nature that they perceived the bear to be. It seems that the most common names for the bear, used by almost all groups of Indigenous

Peoples on both continents, were "Grandfather" and "Grandmother." However, a given tribe or group likely had dozens of different names to choose from. The Cree, for instance, used words with meanings as varied as "son of the chief," "black food," "short tail," "great food," "he who is angry," and "he of great courage" to refer to the bear. Other Indigenous Peoples used names along the lines of "he who must not be spoken to," "great hairy one," and "sticky mouth." The Koyukon People of Alaska, an Athabascan-speaking group, use names such as "the black place," "the dark thing," and simply "the animal." Among the various monikers given to the brown bear by the Finns, one can find "apple of the forest," "pride of the woods," "famous light foot," "blue tail," and "snub nose." The Sámi People call the bear "old man in a fur coat," while in Estonia it is known as "broad foot." The ancient Hungarians—Magyars—borrowed the Slavic term *medve*, thereby avoiding using their own word to refer to the bear. In the lowlands of Kolyma in Siberia, the Yukaghir People referred to the bear as "great man" or "lord of the land." And on the island of Hokkaido in Northern Japan, the Ainu People lauded the bear as the "divine one reigning over the mountains."

One might think that bear hunting would have been a seasonal pursuit, but hibernation didn't deter some hunters. In winter, bears were simply killed in their dens. In the words of the Russian explorer Stepan Krasheninnikov, describing bear hunting in Kamchatka:

> Before the introduction of fire-arms, [hunters] used several devices for killing the bears. Cutting several billets of wood, they stop up the mouth of the den with them, which

> the bear draws in that his passage may not be shut up. This they continue until he is so straightened in his den that he cannot turn himself; then they dig down from above, and kill him with their spears.[4]

Once the hunters returned to the village, the bear they had killed would be greeted with a welcome befitting of a chief. A great feast would ensue and the villagers would make offerings to the bear. Through this ritual, the bear would transform from game animal to revered messenger. Its spirit would appreciate the welcome it was given and reassure other bears, so that they too would allow themselves to be killed by humans and respected in turn.

In the account of his journey to Lapland in 1681, the French playwright and traveler Jean-François Regnard recounts how the Sámi People received a bear once they had killed it:

> When they have killed the bear, they put it in a sled to carry it to a cabin, and the reindeer that pulled the sled is absolved from its duties to pull that sled all season long; they also make sure that buck is kept away from any does. The cabin is built for one sole purpose, and that is to cook the bear; all the hunters then gather here with their wives, who are forbidden to eat any of the creature's posterior parts and are always given its anterior parts instead. The entire day passes in amusement, but one notes that all those who helped to kill the bear are forbidden to approach their wives for three days, following which they must bathe and be purified. [...] There is nothing a Laplander holds in

> greater esteem than to have witnessed the killing of a bear, and it is something he will glorify for the rest of his life.[5]

Three centuries later, Frederica de Laguna, an American ethnologist studying the Indigenous Peoples in Yukon, Canada, recounted a similar story, explaining that

> the dead bear was received like a fallen chief in the hunter's cabin and offered fine clothing; a feast was prepared in its honor, the idea being that it would carry with it the essence of these gifts, and care was taken to exclude the women while the men were feasting.[6]

It was a different time, on a different continent, in a different culture, but the rituals and the respect for the animal were the same. There are many examples of rituals like these spanning the parts of the world where bears are found.

Over the years, hunters have tried every possible method in their quest to kill bears. The classic technique used in ancient times was to kill a bear with a bow or spear. Bait would be set inside a loop of rope tied in a slipknot on the ground, connected to a tree that had been bent over; once the bear sniffed the meat and stepped inside the loop to get it, the rope would tighten, sending the tree springing back upright and taking the bear with it, leaving the bear hanging in midair. Another classic method was to attach bait to a heavy piece of lumber suspended from the rafters of a cabin. As soon as the bear tugged at the bait, the lumber would fall and crush it to death. I've seen this system used on Hokkaido. Other traditional methods include pits with stakes camouflaged under a bed of leaves, and even extra-large spring-loaded leghold traps. Later, hunters learned to poison bears, shoot them with firearms, chase them with dogs, and catch them with a lasso like cowboys in the American West. Fire traps have also been used to kill bears, with devastating results. A method like this can't be described as a traditional hunting technique—instead, it's a tool designed for mass destruction of the species.

In the Pyrenees between 1900 and around 1970, people declared outright war on bears, and the population was decimated, going from approximately four hundred to only around thirty individuals. Using firearms—some more sophisticated than others—local hunters soon inundated the mountain taverns with bear meat and guided plenty of wealthy trophy hunters to rack up impressive kill counts of their own. In the 1940s, Marcel Couturier, a renowned doctor and mountain-game hunter in my home country, called for regulations to be adopted and for certain areas of the Pyrenees to be designated as protected. Couturier self-published *L'Ours brun*, a study of the brown bear, in 1954, and it remains a classic reference today; it's fascinating to delve into the precise anatomical details he provides, and his words are both scientifically grounded and brimming with empathy as he describes the plight of this bear:

> Man has played a role that must not be exaggerated, nor played down, yet is indisputable. Destruction has been systematically practiced under the fallacious pretext that the brown bear is a pest and a dangerous beast. Many wrongs have been committed in this respect.[7]

Strong like a bear

AFTER CENTURIES OF being driven away and witnessing the destruction of its habitat, the bear is perhaps now starting to take revenge. In their prescience, the healers, shamans, and doctors of ancient times realized bears had physiological capacities that could cater to human needs. Our ancestors have always believed that bears hold a universal remedy for a great many ailments. A good dose of the spiritual and mythical connection between bears and humans must have factored into this pharmacopeia. While the early healers had none of the investigative technology we have today, they did have the benefit of thousands of years of empirical knowledge. Today, it's easy to say that many of these traditional remedies were nothing but snake oil, but how do we really know for sure?

The Greek philosopher Theophrastus believed that bear flesh would swell and grow, so eating bear meat would bulk up the human body. This was once recommended as a remedy for anemia.

Bear fat was widely used for cosmetic purposes, to soften the skin or stimulate hair growth. The logic was that bears were very hairy—so their fat must have hair-producing properties! People also used to rub bear fat on bruises and believed it was a remedy for rheumatism and gout. In Finland, it was dried and used to treat poor eyesight, toothache, and ulcers. Meanwhile, in Germany, it was a recommended treatment for swelling.

In the time of the Roman Empire, Galen touted bear bile as a remedy for toothache, and Pliny the Elder recommended its use for gangrene, epilepsy, asthma, and jaundice. Perhaps more plausibly, bear bile has also been used for centuries as a remedy for kidney stones. How did the early healers come up with these ideas? Pharmaceutical science has since demonstrated that the properties of the ursodeoxycholic acid found in bear bile are legitimate.[8] According to Japanese legend, the samurai would take a swig of bear bile before they went to battle. Today a synthetic version of this remedy is used to treat patients with renal colic, as it can dissolve stones in the gallbladder without any side effects. Other anti-inflammatory properties of the acid are thought to be beneficial not only for the liver, but also for the digestive tract and the brain. This may lead to therapeutic applications for degenerative inflammatory diseases, as ursodeoxycholic acid is thought to be able to cross the blood-brain barrier.

Many doctors over time have believed that the animal kingdom holds the solution to ailments in humans. Biomimicry offers a new way forward by tapping into millions of years of natural selection. Instead of conducting long, expensive research studies and obtaining unpredictable results, we can simply observe and study nature to point the way toward a solution. The dividing line between humans and animals may not be as rigid as we might think. Going on the assumption that all mammals—humans included—share a common ancestor, it's not unreasonable to expect them to share similar pathologies. By studying the adaptive responses certain species have developed to extreme environmental conditions, perhaps we may be able to discover new treatments for humans, through an interdisciplinary approach involving doctors, veterinarians, and biologists.

When we were making the documentary *Fort comme un ours* (Strong like a bear) in 2018 for the French network Arte, my colleagues and I spent a year shadowing the team of scientists working with the Scandinavian Brown Bear Research Project. One of the researchers working on the project is Ole Fröbert, a Danish cardiologist at Örebro University Hospital in Sweden who hopes that studying brown bears can lead to the development of innovative treatments for patients with cardiovascular diseases.[9] Keen to push beyond the boundaries of traditional laboratory research, he believes the unique physiology of bears may hold the answer to various medical conditions.

How can bears go for six months without eating or urinating and barely move at all during that time, then emerge in the spring in perfect health, with no signs of muscular atrophy or any other imbalance in their bodies? By contrast, if human beings don't use their muscles, they can expect to lose approximately 10 percent of muscle mass per month of inactivity.

Veterinarians have been working with the Scandinavian Brown Bear Research Project to study brown bears in the Orsa region of central Sweden for around twenty years. Since Ole Fröbert and other doctors and researchers joined the team a decade ago, the project has announced some remarkable findings. Every year in June and February, researchers, mainly from Oslo, Copenhagen, and Stockholm, but also from farther afield—travel to the area to gather the samples they need to conduct their year's work. If an expedition is delayed or canceled due to weather or technical difficulties, this might put as much as a year of research on hold.

Bears fitted with tracking collars are located and tranquilized by specialized veterinarians, including Johanna Painer of Vienna, Austria. In the field, is isn't always easy for the team of veterinarians and scientists to capture the bears. As soon as a bear is tranquilized, around fifteen renowned researchers and students swoop in and start taking the blood, tissue, and fecal matter samples that will be used to provide material for as many as forty different research programs.

The team also gathers information about the bear's health by measuring its height and weight, taking hair samples, and conducting an echocardiogram, among other tests. The whole process takes about an hour, and when they're done, the team leaves the bear at the foot of a tree after taking all the necessary precautions to make sure it wakes up in optimal conditions.

As if exacting an unconscious form of revenge, the bear has become an ideal for humans once again. We've distanced ourselves from the natural world and embraced a sedentary lifestyle, filling our bellies with too much meat and living ever longer, but not necessarily in better health. It's perhaps poetic justice that wild creatures like the bear are coming to the rescue of humans who have grown disconnected from nature.

One of the observations driving Ole Fröbert's research is that a bear can eat up to 250,000 blueberries (*Vaccinium* sp.) a day—about twenty thousand calories worth of fruit, weighing as much as an adult human—without exhibiting any risk factors associated with being overweight. Maybe the remedy we're all looking for lies simply in what we eat. The health benefits of blueberries are many and varied—they're high in fiber, vitamins, and minerals, and low in sugar—with no side effects to worry about.

In fact, the virtues of blueberries have been recognized for hundreds of years. In Europe, for instance, they were once used to reduce diarrhea and dysentery and treat the symptoms of scurvy. Today, we know that blueberries are the king of antioxidant foods, containing vitamin A, C, and E molecules that reduce the effects of aging in the cells of our bodies. These substances are said to help protect against cardiovascular disease and some cancers and to help reduce blood pressure in individuals who are overweight. Clearly, the berry-eating bears are on to something!

At the Hubert Curien Institute in Strasbourg, France, Fabrice Bertile is conducting research into muscular atrophy using the bear samples collected in Sweden. The French National Center for Space Studies is particularly interested in research like this into treatments that might help limit muscle loss in humans, which would be useful in the event that travel to the planet Mars becomes a reality. Not all of the potential applications are so far out there, however. One practical use for this kind of treatment would be to help patients who are confined to a hospital bed for months on end. One might argue that being bedridden is not too dissimilar to hibernation, so who better to look to for a solution than bears?

Human muscle cells exposed to bear serum (the clear liquid separated from blood) collected in winter have been found to react favorably, by not only growing in volume and contractile strength but also deteriorating less rapidly. Exposure to the bear serum was also found to promote an increase in fat consumption, thereby limiting weight gain. Researchers are exploring the theory that hibernating bear serum can help conserve human muscle mass and promote optimal use of body fat. If they can isolate the molecules responsible for these effects in bears, they may have found a solution to the problem of muscle wastage in astronauts and individuals whose physical activity is limited.

Alongside these programs, at the bear research center in Orsa, scientists are studying bears in captivity by using non-invasive techniques and basic anesthesia. Peter Stenvinkel is a professor specializing in renal (kidney) medicine at the Karolinska Institute in Stockholm, Sweden.[10] Like osteoporosis, kidney disease causes bone fragility that can lead to an increased risk of fractures. Bone fragility impacts as many as one in three women around the world, and men are also at risk. In Sweden, where much of the population is aging and the streets in the winter are covered with ice and snow, there are five times more deaths from pedestrian falls than from automobile accidents. Bears, though, exhibit no signs of osteoporosis, despite not moving for as long as six months during the winter. Through his research work with bone specialist Mathias Haarhaus and radiologist Torkel Brismar, Stenvinkel is trying to identify the mechanisms in the brown bear's body that prevent the loss of bone mass during these long periods of winter inactivity. It could have to do with kidney function: during hibernation, blood flow to a bear's kidneys is reduced by 90 percent, yet its kidneys continue to function normally.

These researchers are using advanced technology, including 3D scanners and precise analysis of fatty acids, to study the significant variations in the bear's body between its two major periods of the year: hibernation and activity. Areas of bone growth can be clearly observed during the summer but not in the winter, and the concentration of certain molecules in the blood varies similarly. Essentially, researchers are trying to isolate the substance that prevents bone degeneration in bears.

Meanwhile, other studies are underway to find out more about the changes that occur in the bear's intestinal flora. There's plenty of talk today about the impact of the gut microbiota on human health. We know now that our health depends on the billions of bacteria, cultures, and viruses that live in our digestive tract in symbiosis with the rest of our body. In order for us to adapt to changes to our diet, environment, or climate, it's important that our body and our gut microbiota continue to work together.

Perhaps there are new insights for human gut health to be gained from studying the variations between the bear's feeding and hibernation periods. As bears prepare to retreat into their dens, they ingest dirt from their surroundings through the food they eat. Is it possible that the bacteria in their hibernating environment may bring about a change in their intestinal flora?

Scientists at the University of Gothenburg are also making use of the samples gathered from the bears of Ursa. Their research has indicated how the bear's intestinal flora varies in quantity and effectiveness from summer to winter. Dr. Fredrik Bäckhed's research has involved transplanting microbiota into the gut of germ-free mice, then rapidly increasing the body weight of a group of mice that were given the microbiota and a control group of mice.[11] All of the mice became obese, but those with the bear bacteria didn't develop any complications. Dr. Bäckhed's colleague Felix Sommer also observed that the adipocytes (cells that store fat) in the mice with the microbiota were more effective at storing fat. And in an earlier study, Dr. Bäckhed discovered that bear microbiota composition can influence the quantity of energy harvested from food.[12] These findings point the way to promising avenues of research that may increase our understanding of human obesity.

Across the Atlantic, Dr. Ron Korstanje at the University of Maine is going one step further by studying the bear genome. His team has succeeded in sequencing the genome of the black bear at different times of the year. When comparing the results, these researchers observed that some genes are active in summer and not in winter, which would suggest that bears are able to put part of their metabolism on hold depending on the season. It's thought that some of their genes serve specifically to regenerate kidney function following hibernation. Now researchers are trying to determine which genes act in this way and in what proportion. They hope that once they've figured this out, a solution can be developed to provide an alternative treatment to transplant and dialysis for some 500 million people around the world suffering from kidney disease and failure.

While bears may not hold the immediate solutions to the health problems affecting millions of humans around the world, they can open new avenues for research into a variety of conditions, including osteoporosis, obesity, muscular atrophy, renal and cardiovascular diseases, neurological disorders, and degenerative brain diseases. We might think of the insight they promise as a tool kit that has taken millions of years of evolution to put together.

As we can see, bears are finding a way to get their revenge as this new research brings our species and theirs closer together again, and makes us realize how much we need them. It was about time they fought back against the popular beliefs and religious precepts that drove our ancestors to try to wipe them out. Perhaps these research initiatives will serve as a wake-up call and help us to see how closely connected we really are to the other living beings around us.

How cave bears succumbed to bone disease

At one time, cave bears (*Ursus spelaeus*) roamed an area that covered France and the south of England and stretched as far east as Russia. The earliest evidence of the species can be traced back around 250,000 years. The cave bear was a descendant of Deninger's bear (*Ursus deningeri*), which lived in the same territory from about 1.7 million years ago until it died out, approximately 100,000 years ago. During this time, brown bears were scarce.

Though the last cave bears are thought to have disappeared around 10,000 years ago, there would have been a mass wave of extinction between 35,000 and 24,000 years ago, at the peak of the last ice age.

For a long time, cave bears were thought to have been exclusively herbivores, but studies of the microwear on their teeth have since revealed that they ate a mixed diet of plant foods and meat. There are various theories as to why cave bears rapidly disappeared, and one suggestion is that they may have starved to death because the plants they were used to eating could no longer survive as climatic conditions grew colder. Another theory suggests that intensive hunting by humans may have led to their demise.

I'd like to venture another possibility. Obviously, the disappearance of the species is due to multiple causes, but one of these may not have been considered before now: What if cave bears simply weren't as well adapted for hibernation as brown bears?

Some serious consideration might be given to this theory in light of the work of the researchers in Stockholm, which has shown how brown bears are specifically able to adapt in order to avoid losing bone mass during hibernation.

Remains of thousands of cave bears thought to have died during the winter have been found in caves across Europe, most notably in Austria, Romania, France, and Germany. As far back as 1837, scientific claims were made that cave bear remains found in a cave in Iserlohn, in the North Rhine-Westphalia region of Germany, showed evidence of bone disease, including a cancerous tumor. One of the largest number of cave bear skeletons was found in the grotte de la Balme à Collomb cave, in the Savoie region of France. Most of these bears died during the winter, and all of them suffered from bone-related disorders such as dwarfism, bone damage or apposition, deforming arthrosis, infectious or inflammatory conditions, poorly healed fractures, or proliferations of cells that had led to protrusions or tumors. In many cases, fused vertebrae were found. And another bony part of the cave bear's body, the baculum, or penile bone, was also often found to have broken.

In the Chauvet Cave, thousands of bone fragments were found to have come from around two hundred different bears that had died over a long period of time. Surprisingly, none of them would have been of reproductive age. But perhaps what's most interesting to note here is that DNA analysis has demonstrated that these cave bears had a low effective population size and limited genetic diversity. Ultimately, these bears in the South of France were close to extinction when early *Homo sapiens* started to draw pictures of them on the walls of their caves.

As temperatures dropped in the lead-up to the last ice age, snowfalls must have become more abundant and cave bears would have been forced to spend more time overwintering in a cavern. Their capacity to handle a long period of fasting and immobility was probably far from optimal.

It has also been suggested that the absence of light deep inside a cave for long periods of time would have deprived these bears of the vitamin D their bodies needed for proper bone growth. As we know now, hibernating brown bears don't suffer from this problem.

Bear skull in Chauvet–Pont d'Arc

CHAPTER 8

Bears and Humans—the Future

*And each generation in turn will ask:
Where is the big white bear? It will be a
sorry answer to say he went under while
conservationists weren't looking.*

Aldo Leopold[1]

THE RELATIONSHIP THAT has connected humans and bears for the last 500,000 years has changed a lot since the time of the Neanderthal hunter-gatherer. The absence of bears from the *Physiologus*, the first Christian book of beasts, written in Egypt around the second century CE, is perhaps an early indicator of humans' willingness to write bears out of the story. It seems surprising that bears wouldn't figure in this book, as they were widespread across nearby countries such as Syria and Greece at that time.

Up to the Middle Ages, three animals were thought to be closely related to humans. Pigs were valued because their anatomy was similar to our own (this made them well suited for medical research, since dissection of the human body was forbidden by the Church); monkeys, because they imitated us, were thought of as more devious versions of ourselves; and bears were thought to resemble humans in their stature and behavior. Bears were seen as a dirtier and more boorish version of ourselves—the dark, malodorous side of humankind.

Christianity might have tried its damnedest to knock the bear from its pedestal, starting as early as the fifth century. But in spite of the fragmentation of bear populations caused by hunting and the increasing human dominance of their habitat since the Neolithic era, a certain respect for these creatures remained anchored in the hearts and minds of the Indigenous Peoples who lived in their midst, as it had since the dawn of time.

Starting in the seventeenth century, though, the outlook for bears grew bleaker as attitudes began to change. As historian Éric Baratay has pointed out, there was a shift toward anthropocentrism—the idea that humans are the most important entities in the universe—under the influence of the Bible and clerical discourse during the seventeenth and eighteenth centuries:

> In 1644, the priest Louis Bail wrote with respect to animal species: "Some can be ridden or used to pull a carriage; others serve for pleasure, such as game animals; others provide ivory for ornament, or wool or silk for clothing;

> others provide meat from their own flesh; while others serve medicinal and other purposes."[2]

As for hunting, this was justified as a way of affirming the dominance of mankind. In fact, the systematic culling of wild animals was seen as a duty. As Baratay has further clarified:

> Hunting was the primary form of interaction between humans and these animals as it was considered a necessary means to restore order. The Capuchin theologist Yves de Paris, for example, saw hunting as a way for mankind to assert its power by triumphing over a strong, cunning animal after a bout of light combat of bodies and minds. But this was not the only purpose it served. In 1613, the priest Louis Gruau cited five justifications for hunting: "to repel deleterious beasts and procure pelts and victuals, but also ward off idleness, awaken numb bodies, and ruses of war." Hunting thus introduces the elements of distraction and satisfaction of curiosity to mankind's rapport with wild animals.

Two centuries later, bear hunts were being portrayed in some venerable nineteenth-century texts with lyrical flights of fancy tinged with a certain compassion for the bear. As the French aristocrat and writer the Marquis de Cherville wrote about the plight of bears at the time:

> The hatchet has stripped the guardian summits of the Alps and the Pyrenees; man, ill at ease in the plains, has invaded the mysterious depths of the solitude protecting the species from destruction; the bear has become rarer than rare. [. . .] I had always held the venerable remnants

> of these early animals in a certain esteem; I was filled with compassion for the cruel destiny and humiliation civilization has reserved for them in their ancestral forests; my sympathy was heightened with some respect for the philosophical character I believed I would discover in their love of solitude.[3]

Can the Marquis's rhapsodic display of compassion make us forget his relentless hunting? Certainly not. As protests against hunting have illustrated, the polished language of many an aristocrat can quickly lose its shine when their feudal recreation comes under threat.

There was a special place reserved in the bestiary of hunting for the game of princes and kings. Even so, these hunters were aware of the increasing scarcity of big game and had a deep knowledge and understanding of their habitat. This connection was broken, however, as bear culls gained momentum across Europe and the United States. Once considered noble game, the bear had now become a nuisance. An article published in a popular magazine in 1908, entitled "In the Bear's Claws," illustrates the image of the bear that had taken shape in the public mind. The introduction reads as follows:

> Of all the big game hunts, this is one of the most stirring. Cruel and fearsomely armed, the Lord of the forest takes a great many lives each year. What is more, the lure of danger sparks the passion of certain amateurs of the sport, be it in the vast glacial forests of Russia, or the mountains of the New World where the savage grizzly has not finished spreading terror.

While this introduction may seem to justify a lawful eradication of the population, the conclusion, tinged with both regret and relief, offers something of a cautionary tale:

> The war we wage on them is merciless. In many countries, rifle hunting is supplemented by traps and poisoned bait. Each year, land clearing and deforestation are encroaching further and further on the kingdom of the ferocious master of the forests. It is foreseeable that, in forty years perhaps, the bear will have become a rarity, to the great despair of unrepentant hunters and lovers of excitement, and the rejoicing of peasants who live in fear of the beast to this day.

People had their reasons for killing bears, be it in the name of sport for the wealthy or opening up new lands for peasants to cultivate. These justifications for killing the bear were not dissimilar to those cited in works as early as Gaston Phoebus's fourteenth-century masterpiece, the *Livre de chasse*:

> Bears are of two conditions: some large in nature, and others small in nature, even if they are old. However, their morals, lives, and forms are identical, though the larger ones are stronger and sometimes feast on deprived animals. They are marvelously strong in all their body, except their head, so weak; if they are hit, they are easily dazed, and if they are hit hard, they shall die.

Bear hunting today has none of the nobility or ethical spirit that inspired these writings. All the respect for the animal has gone. Long-range sniper rifles, exploding bullets, and helicopters are now the weapons of choice, and wealthy clients simply pull the trigger on the bear from a safe distance, under the orders of their guide. Even the age-old biblical injunction to affirm human dominance can't justify today's methods.

Jim Harrison, the great American author and lover of wild nature, sums up the disgrace of modern bear hunting in a few short lines:

> There is the usual tale of the grizzly hunt where we are led to assume that the bear had spent its entire life hell-bent on murdering the author, rather than merely walking around in the woods looking for lunch. And no matter that the animal is shot at 200 yards, before it can see the hunter. I can remember an account [from a hunter] where the grizzly was asleep, something to the effect that "I poured hot lead into Mister Dozin' Bruin. It was the surprise of his life!"[4]

Bear hunting is a pursuit revered by esteemed sportsmen, idle bigwigs, and billionaires, whose sole aim is to snap a picture of their trophy so they can show off on social media how brave they were to kill such a savage beast. Romain Gary said it all in his novel about the slaughter of elephants, *The Roots of Heaven*:

> Lovers of a nice shot had flocked there from all over the world, certain of getting their money's worth: over fifty safaris in one month, a fine congregation of impotents, alcoholics and those females whose sexuality is so pleasantly tickled by bullfights and who reach their climax with a finger on the trigger and an eye fixed on the horn of a rhino, or the tusk of a fine male.[5]

There are companies in Alaska and on the Kamchatka Peninsula that specialize in this kind of hunting trip. The hunters mainly target fine breeding males in the prime of life that have fought tooth and nail to rise through the bear hierarchy—only to now be felled with the pull of a trigger from afar, while they were simply munching on blueberries. On Baffin Island in northern Canada, I once came across a French hunter who had just killed a polar bear. Because the hide was smaller than promised in the contract he had signed with the company, the guide was taking him black bear hunting to make up for it. The conversation came to an abrupt halt right there. He might have been a fellow countryman of mine, but we had absolutely nothing to say to each other. The hunters always say there are too many bears, so they have to be killed to regulate the population. As if anything could truly justify this sick behavior. And when all is said and done, the hide of a fine adult male bear will make a plush rug or wall hanging in the home of a hunter whose only virtue was to spend tens of thousands of dollars for his own pleasure.

I love rereading the works of the great philosophers of nature, those who shaped our vision of a new relationship to wild animals and vast open spaces. Among them is John Muir, one of the pioneers to explore Alaska and the American West, and cofounder of the Sierra Club conservation organization. In his book *A Thousand-Mile Walk to the Gulf*,[6] he makes it clear that he is on the bears' side. In a country barely unified by the American Civil War, he cites the Bible and offers an alternative interpretation:

> To me it appeared as "d[amne]dest" work to slaughter God's cattle for sport. "They were made for us," say these self-approving preachers; "for our food, our recreation, or other uses not yet discovered." As truthfully we might say on behalf of a bear, when he deals successfully with an unfortunate hunter, "Men and other bipeds were made for bears, and thanks be to God for claws and teeth so long." Let a Christian hunter go to the Lord's woods and kill his well-kept beasts, or wild Indians, and it is well; but let an enterprising specimen of these proper, predestined victims go to houses and kill the most worthless person of the vertical godlike killers,—oh! that is horribly unorthodox, and on the part of the Indians atrocious murder! Well, I have precious little sympathy for the selfish propriety of civilized man, and if a war of races should occur between the wild beasts and Lord Man, I would be tempted to sympathize with the bears.

The dawn of a new relationship

CONTEMPORARY HUMAN-BEAR RELATIONS are contrasting, if not paradoxical. Who doesn't remember cuddling their teddy bear and listening to the story of *Goldilocks and the Three Bears* as a child? Generations of children have fallen in love with bears in stories and animated movies. Among them are A.A. Milne's *Winnie-the-Pooh*, published in 1926 and followed by a Disney animated movie in 1977, Michael Bond's *A Bear Called Paddington*—a spectacled bear, as it happens—which was published in 1958 and brought to the big screen in

2014, and Ernest the bear in Gabrielle Vincent's 1981 picture book *Ernest and Celestine*, which was made into an animated movie in 2012. Endearing as they may be in the pages of a picture book or on a movie screen, bears are still wild animals. Parents are not afraid to remind their children of this, but can only take them to zoos to see the depressed, obese specimens kept in captivity.

When the teddy bear was developed by toymakers in the early 1900s, it heralded a new pact between humans and bears. Two different companies, one in the United States and the other in Germany, claim to have come up with the idea independently of each other. As the story goes, in 1902, President Theodore "Teddy" Roosevelt, who was a keen bear hunter, spared a bear cub from death after he had been invited to shoot it so as to not go home empty-handed. He refused to pull the trigger, and the scene was satirized in a political cartoon in a major newspaper. This inspired a Russian-born toymaker by the name of Morris Michtom to create a soft toy bear and send it to Roosevelt. He put another plush bear in his shop window and nicknamed it "Teddy's bear"—and the rest is history. (In my native France, people soon gave these soft toy bears a nickname of their own—"Michka," an old Russian pseudonym for the bear.) At the same time, Steiff, a toymaker in southern Germany, developed a toy for children in the form of an articulated stuffed bear. It didn't take much for people to adopt these adorable little bears, and a new affinity between humans and bears was born.

Regardless of who invented the plush bear we've come to know and love, the toy companies rose to the occasion to meet what had quickly become a societal need in the early twentieth century. Psychiatrists have studied the visceral attachment that forms between children and their stuffed toy companions. Their teddy becomes a friend, a confidant, the perfect transitional object that can act as a substitute for their mother. A teddy is soft and comforting, and as it is rarely, if ever, washed, its smell can be very reassuring to a child. Losing or misplacing a teddy bear can be a highly upsetting experience for a child as it may feel like being separated from their mother. A blankie can be any kind of animal or even just a piece of fabric, but a teddy can only be a bear.

Among the latest cinematographic representations of teddy bears are *Ted* (2012) and its sequel, *Ted 2* (2015), featuring a plush bear who talks and acts like a human being. In a way, these movies hearken to the themes of the ancient legends: the ultra-virile bear with a taste for the good life; the human double who transgresses the rules. Ted engages in drugs, profanity, and sex, all with the appeal of a cute, fluffy teddy bear—but one that is clearly a reflection of his human companions.

Many readers will be familiar with Smokey Bear, the stylized mascot for the US National Park Service's wildfire prevention campaign who made his first appearance on a campaign poster on August 9, 1944. Not many people, however, may know that in the spring of 1950, a black bear cub that was miraculously saved by firefighters from a fire in the Capitan Mountains of New Mexico became the real-life mascot of the campaign. The cub had suffered burns to its paws and went on to live until 1976 at the zoo in Washington, DC. For decades, the image of Smokey Bear has warned visitors to US National Parks about the dangers of forest fires. Smokey's message is a simple one: THINK.

Meanwhile, grizzlies were being culled by the dozen in Yellowstone Park. For decades, they had been free to roam around campgrounds and feast on the garbage campers left behind. Mothers would bring their cubs, and the cubs would then return on their own, and so on. Why wouldn't they, since the food was easy pickings? However, when the authorities suddenly decided to cut off their access to the garbage dumps, a string of bear attacks on visitors ensued. Grizzlies were systematically culled to solve the problem, and the bear population plummeted to just 135 in 1976. Now the Yellowstone grizzly is classified as an endangered species. Thanks to the protective measures put in place, the population had risen to 690 by 2016. In 2017, authorities stripped the species of its threatened status, though in 2018 it was ordered to be restored by a US judge. This ruling was justified in part because the growing grizzly population in Yellowstone is on its way to interbreeding with other populations to the north, which will improve the genetic diversity and long-term health of the species. But with bears, nothing can ever be taken for granted. There were hopes that the Smokey Bear campaign had heralded a new era of humans embracing nature, the breach of a philosophical divide. As we can see, however, it's a fragile situation.

Outside the realm of movies for children, bears are still generally portrayed as dangerous predators that will savagely attack humans. Just some examples of this are the

2015 Hollywood blockbuster *The Revenant*; *The Edge*, a 1997 thriller starring Anthony Hopkins; Jean-Jacques Annaud's 1988 drama *The Bear*; *Into the Grizzly Maze*, a 2015 movie in which campers are stalked by a man-eating bear; and the 1971 Western film *Man in the Wilderness*. All of these movies were filmed using bear actors—huge bears raised in captivity that are twice the size they would be in the wild—or animatronic bears, in the case of *The Revenant*. These staged movie scenes only perpetuate the terrifying vision we've been persuaded to believe is true about bears in the wild, just as movies about great white sharks have done. Once again, bears have been demonized, representing everything humans can't understand or control themselves.

Nonetheless, brown bear attacks on humans are rare. When they do happen, it tends to be in situations where the bear is caught by surprise and lashes out at the intruder, usually at their head. From 2010 to the time of writing these words in early 2020, eighteen fatal attacks by brown bears were recorded in North America, plus around ten fatalities due to black bear attacks.

Bear attacks in captivity are rare as well, though they do happen. The atmosphere was tense when I traveled to Sweden's Orsa Rovdjurspark in August 2017 to film scenes for a documentary. The previous week, a young staff member had been mauled to death by a two-year-old brown bear in an unfortunate incident that highlighted a deficiency in safety precautions. The nineteen-year-old intern had been training to be a zookeeper, living a dream by working with these large predators. The incident occurred as zookeepers were preparing to allow guests into a protected area between two enclosures for a special up-close experience, but one of the bears managed to dig its way under a fence into the safe zone. Caught by surprise, the bear attacked and seriously injured the intern, who later died of his injuries. The bear was euthanized.

Bears, and especially solitary individuals and females, enjoy their peace and quiet—just like many of us humans. Certainly, when we're in bear territory, we must be on our guard. However, we should remember that bears will often tend to avoid us rather than seek us out.

On my desk there is a black-and-white photo of a brown bear in the midst of a steaming hot spring, a gift from the Russian natural scientist and photographer Vitaly Nikolayenko on my first visit to the Kamchatka Peninsula in June 1993. He showed me around his office on the mezzanine of a magnificent wood cabin with sweeping views over the Valley of Geysers. All his notes and photos he had taken of bears were neatly organized in this den of his. He told me all about his experiences bathing in the hot springs with bears standing mere feet away, and explained that since they were very close neighbors, he typically recorded as many as eight hundred sightings every year. A friendly soul with a twinkle in his eye, Vitaly seemed to live for these bear encounters. Some years later, his cabin burned to the ground, along with a lifetime of work and memories. Was it an act of arson? Maybe. Vitaly wasn't home and his cabin was getting in the way of tourist development in the area. Also, his stance against poachers had not made him a popular man. In December 2003, Vitaly skied after a bear that should have been in its den for the winter. But he got too close. The bear didn't tolerate his presence, and it mauled him to death.

This man had spent thirty-three years in the midst of brown bears, studying them and trying to unlock their secrets. He was a controversial character because of his intrusive approach toward observing bear behavior. He was aware of the potential risks but thought he was tuned in well enough to bears to be able to understand them and socialize with them.

Timothy Treadwell was an American bear enthusiast who thought the same thing. In October 2003, he and his girlfriend were killed by a bear. He had been approaching bears for thirteen years, inching closer every time, and never carried a weapon.

As I've said already, if you know what a bear is going to do, you know more than the bear does. All you can do is stay humble and remember never to kid yourself that you can read a bear's mind.

Yes, these big carnivores are always unpredictable, but these examples also illustrate that as a general rule, bears don't always represent a threat. Certain individuals, in certain circumstances, can be dangerous. One might argue that the same goes for humans.

In his description of the Kamchatka Peninsula published in English in 1764, Stepan Krasheninnikov wrote:

> The bears of *Kamtschatka* [*sic*] are neither large, nor fierce, and never fall upon people, unless they find them asleep; and then they seldom kill any one outright, but most commonly tear the scalp from the back part of the head; and,

> when fiercer than ordinary, tear off some of the fleshy parts, but never eat them. The people who have been thus wounded, are called *Dranki*, and are frequently to be met with. It is remarked here, that the bears never hurt women; but, in the summer, go about with them like tame animals, especially when they gather berries. Sometimes, indeed, the bears eat up the berries which the women have gathered, and this is the only injury they do them.[7]

This vision of the interaction between bears and women has proven to be somewhat idyllic, as anthropologist Nastassja Martin, the survivor of a Kamchatka brown bear attack in August of 2015, can attest. There are plenty of lessons to be learned from her experience, which resonates especially deeply as Nastassja has studied the place bears hold in the legends and dreams of the Indigenous Peoples of Alaska and Kamchatka. In her account of the encounter, she writes:

> Yet, it is at the heart of the glaciers, in the midst of the volcanoes, far from people, trees, and salmon, that I find him, or he finds me. I've been walking on this lofty, arid plateau, a place where I have no business being, really. As I step off the glacier, I make my way down from the volcano. Behind me, the rising smoke forms a halo of clouds. I think I'm alone, but I'm not. A bear just as disoriented as I am is wandering these same heights where he has no business being either. He looks like a mountain climber, almost. What's he doing here, in this barren land, with no berries or fish in sight, while he could be lazily fishing for his next meal down in the forest? Our paths meet. This is what they call

kairos. If ever there were a critical moment, this is it. A hump on the landscape is hiding us from the other's view, the mist is rising, and the wind is blowing the wrong way. When I see him, he's right in front of me. He's as surprised as I am. We're only a few steps apart. There's no way out—for him or for me. I remember what Daria told me: "If you come across a bear, tell him 'I'm not out to get you, and you're not out to get me either.'" Yes, well, that's not going to work this time. He bares his teeth, he must be scared. I'm scared too, of course, but I can't run away, so I do the same thing. I bare my teeth at him. After that, everything happens very quickly. We collide, he knocks me down, my hands are in his fur, he bites my face, then my head, I can feel my bones cracking, I tell myself I'm dying, but I don't die. I'm fully conscious. He lets go and grabs hold of my leg. This is my chance. I reach for my ice axe—it's still in my holster from my walk on the glacier this morning—and I hit him with it. I don't know where, since my eyes are shut tight. I'm not seeing, only feeling. He lets go, I open my eyes, I see him running away with a limp, I can see there's blood on my makeshift weapon. And I just lie there, stunned and bleeding, wondering if I'm going to live, and I do. I'm more lucid than ever, my mind is racing a hundred miles an hour. I tell myself, if I live, I'm going to have plenty to say and do in this life of mine. I tell myself, if I live, this will be my second birth.[8]

In the relatively short list of bear attacks, the case of Michio Hoshino stands out. One of the most talented wildlife photographers of his generation, he had expert knowledge of the animals native to Alaska: grizzlies, elk, and black bears were his subjects of choice. But on August 8, 1996, this discreet, respectful photographer was killed by a bear on the Kamchatka Peninsula, around the Kuril Lake area. The bear dragged him out of his tent and took him into a bush to eat him, in a very rare case of predation.

As we've seen with bears, their dangerousness is relative. Fatal encounters with bears happen, yet they are very rare.

When the bear population was significantly higher and more humans lived in the woods and alpine pastures, incidents like these must have been more frequent. The conflict between cattle farmers and large predators isn't new. It was generally accepted that around 5 percent of a herder's flock might be lost to bears or wolves, and the same is still true of yak herders in Tibet who have learned to live with the snow leopard in their midst. It's the price to pay for occupying their territory.

The state of bears around the world

THERE'S A STARK contrast between bear populations in different parts of the world. Exact numbers are hard to determine, but the country with the greatest population of bears is undeniably Russia. The brown bear is Russia's national animal, and its territory extends from the country's eastern border with Europe all the way to the Pacific coast. Its population is estimated to be around 120,000 individuals, and some 15 percent of these can be found on the Kamchatka Peninsula.

In the contiguous United States (not including Alaska), only 1 percent of the population that existed before human settlement remains today—around 33,000 bears.

The total bear population in Europe is estimated to be about 17,000 to 18,000 individuals, dispersed across twenty-two countries. In the European Union, brown bears are protected under the Habitats Directive of 1992, which ensures the conservation of their habitat, while the Bern Convention on the Conservation of European Wildlife and Natural Habitats, presented for signature in 1979, covers the bears themselves. Some exceptions to these measures may be granted, such as in Finland, where local authorities can decide to euthanize a bear in the event of an attack on people. In Trento, Italy, a bear was killed in August 2017, not by hunters for its hide, but by an overdose of tranquilizer while it was being transported.

In Romania the bear population has skyrocketed and, according to the local news media, conflicts are becoming more frequent. In late October, bears can still be found sniffing around apartment buildings; climate warming is doing nothing to encourage bears to retire to their dens for the winter at this time of year. Instead, they remain on the lookout for easy sources of food near humans. If they're encroaching on neighboring towns and cities, it's because humans keep building closer and closer to the Carpathian Mountains. Bears don't know where to hide anymore, or what to eat, so they try their luck.

It's a similar story on Russia's Sakhalin Island, where severe drought has resulted in limited resources available in the wild.

In September 2017, two people were killed here, and eighty bears were culled in retaliation.

In Scandinavia specifically, estimates suggest there are around 2,800 bears in Sweden, and just 125 in Norway. Bears, as we've seen before, are opportunistic creatures. They can live in proximity to humans, but they need their peace and quiet in the winter to be able to stay in an area.

France is a cultural exception—as it often is—in the way it deals with large predators. Today, bears and wolves benefit from protected status in Europe and elsewhere in the world. There are strict rules and quotas regarding their killing. But the bear population in my home country has declined significantly since the French Revolution. Hunting was encouraged in order to eradicate a species that was considered a pest. Bounties were offered to hunters, like they have been in recent years for foxes—as if we had never learned from the errors of our ways.

In the Pyrenees, a mountainous region that has been widely populated by humans since the Neolithic era, bears have been progressively driven into the Ossau and Aspe valleys, where they retreat to the steepest, craggiest terrain during the daytime. Bears are usually diurnal animals, but in this mountain range, they tend to become active later, between around eight and ten in the evening. An increase in activity has also been observed in the morning, as this is the time when they move away from their nocturnal activity zones. Around here, one bear will typically kill three or four sheep every year, which means they eat a far less carnivorous diet than many humans. Great Pyrenees dogs do a remarkably good job of limiting predation, and electric fences keep the sheep even safer. Double fences can also help to keep sheep from panicking.

Bears were hunted for grease and captured for trainers in the Pyrenees, and this has decimated a once-flourishing population. (Bear cubs were also fattened up for butchers' shops to provide grease.) There were only around fifteen bears left in the region by 1982, and this number dropped even further in the mid-1990s—to just five—before new female bears from Slovenia were introduced to the area. The original stock died out in 2004 when Cannelle, the last female bear from the Pyrenees, was killed.

In the absence of bears, however, farming practices have flourished, aided by government funding. In one small community in the Pyrenees in 2005, while the government at the time was working to reintroduce bears in the region, the conservation association that had lobbied for the bears' return organized a festival, complete with information booths to help people young and old understand the initiative. But that wasn't the way the local cattle farmers saw it. They sprayed graffiti with the words "Kill the bears" on the roads and vandalized the festival tents at night; guards patrolled the site, and the atmosphere was decidedly unpleasant for any visitors who dared to come. Threats were made against the local mayors as they welcomed the association and its guests. The following week, in another village nearby, where the mayor was keen for bears to be reintroduced as they would represent an attraction for his region, the farmers' anger was even more palpable, as they hurled racist and homophobic insults at the MC of the event. The farmers felled trees over power lines and roads leading into the village. I had been invited to speak at this event, and I found myself speaking to 150 people crammed into a small room. In the middle of my presentation about the coexistence of humans and bears in various countries around the world, from Slovenia to Japan, the farmers started to bang sticks against the floor tiles, spreading panic around the room, and people started to leave. I carried on regardless, showing a series of slides of a mother with four cubs, and one plucky farmer shouted, "Why don't you show me six? I've got six rounds in my shotgun."

Clearly, the anti-bear lobby was still going strong in this part of the world.

By 2016, the bear population in the Pyrenees had risen to thirty-nine individuals, and to forty-three by the end of 2017. The damage they've caused has not worsened, save for one incident in which a flock of sheep fell from a cliff. A bear was suspected of having scared them to their deaths. In August 2017, farmers assaulted a group of officials who had come to investigate a series of alleged bear attacks on their flocks. The farmers insisted that they should be the ones to impose their own justice. This handful of radical activists made threats and fired warning shots in a show of decidedly anti-democratic behavior.

There's just no changing the mentality of some farmers. After thirty years or more, they still haven't come around to the idea that perhaps the mountain herders' business model has to evolve like the rest of the economy and the industry have. As geography professor Farid Benhammou argues, bears are an internal geopolitical issue:

International issues, oil, the Middle East, Iraq—these are the kinds of things we tend to associate with the word "geopolitical." However, I would argue that the term can extend to describe environmental conflicts and rivalries between different groups closer to home. The issue of bears in the Pyrenees is a case in point.[9]

In 2017, of the 570,000 sheep in herds in the Pyrenees, 18,000 to 30,000 deaths occurred due to falls, attacks by stray dogs, or sickness. Even counting the nearly 200 sheep that fell from the cliff, about 500 deaths were attributed to bears—under 2 percent of the total death toll.[10] When flocks are guarded, the risk of loss is negligible. Bears and wolves have been scapegoated to appease the discomfort of mountain dwellers who are stuck between modernizing their practices and wallowing in a pseudo-traditional way of life.

The bear is caught in the middle of a power struggle. Two groups unrelated by political affiliation are clashing heads. On one side, local politicians and people in agriculture are pushing for bears to be completely eradicated from their land, fueled by a lobby of reactionary hunters and sheep farmers. And on the other, environmentalists and concerned citizens are convinced that free-roaming bears in the wild are a sign of the ecological health of their country. The rift between them has grown wider as each successive government has failed to negotiate a solution to the issue that works for both sides—a solution whereby people with one conviction can live beside the other without trading insults, and without dwelling on tradition.

On March 26, 2018, Nicolas Hulot, France's minister for ecological transition, announced that two female bears would be reintroduced to the Pyrenees-Atlantiques region in the fall of that year. These were his words:

> I will be asking the regional authority to manage the dialogue to ensure the success of this reintroduction, and I will personally be there to witness the event. [...] I have decided to take action because there are now only two males remaining in this region, one being Cannellito, whose mother was Cannelle, the last bear of pure Pyrenean stock. I will not be the minister who stands idly by and watches this lineage die out. [...] I refuse to let this species succumb to the same fate as the ibex here in the Pyrenees, or the monk seal in Corsica. The bear is a part of the natural heritage of our country.

Local associations for the protection of bears, along with concerned citizens, had been waiting for this announcement for years. According to a February 2018 survey, 84 percent of people across France were in favor of maintaining a bear population in the Pyrenees. Respondents in the Hautes-Pyrenees region were 70 percent in favor, and those in the Pyrenees-Atlantiques, 78 percent.

I'm not surprised that the government has taken these measures. The minister and I conversed at length during our downtime while we were filming the TV nature documentary *Le Repaire de la licorne* (The unicorn's den) in 2005, and I saw for myself how committed he was to protecting biodiversity, even though this stance exposed him to more criticism than praise. You can't stand up for your beliefs in the face of power without a firm commitment to making change happen.

History repeats itself, and once again bears are crystallizing people's fears and concerns. This time, though, priests are not the scaremongers; rather, it's local politicians and industry leaders who are leading the charge. Sheep farmers have taken advantage of the absence of bears from their land for several decades to diversify their business with lowland agriculture, driven by national and European policy initiatives. Meanwhile, they've left their flocks to fend for themselves against stray dogs, thunderstorms, and other potential hazards in the alpine pastures. Sheep farming in this region is subsidized because of fears it may be driven out by competition from New Zealand. Bears are not the problem. Globalization is the problem. Bears can actually represent a positive influence in some regions, as they demonstrate the conservation efforts being made in the mountains. Bears are not only a draw for tourists; they're an emblem for citizens to be proud of.

Bears know no borders; they are citizens of the world. We've seen how humans and bears in various countries have lived in proximity to one another at different times and in different circumstances. If there's one thing the examples in this chapter can tell us, it's that anything is possible.

When I was in Japan in 2005 to film the documentary *Les Ours du soleil levant* (Bears of the rising sun), I had a memorable conversation with the owner of a salmon fishery on Hokkaido's Nemuro Peninsula. As I've explained, the Hokkaido or Ussuri brown bear—*Ursus arctos yesoensis*—is a distinct and highly populous subspecies in this part of the world. The owner of the fishery and his staff have lived alongside the bears for more than thirty years. The bears wander across the site, seemingly indifferent to the human activity there. This is a place where bears and humans both help themselves to the riches of the salmon spawning season. This is a place where both bears and humans are at home.

I asked the owner how people asserted themselves amid the bears, and this is what he said:

> When we find ourselves right beside a bear, maybe two or three meters [6 to 10 ft] away, we have to stare it down. Once the bear realizes it has lost, it will walk away. That being said, we must never let bears go into the houses, or they will come back and think they can just help themselves there too.

He went on to say that it was important for people there to insist that this was their home, but to share the space mindfully in order to ensure that bears didn't become too habituated to humans and have to be put down.

All around the Nemuro Peninsula, there are little fishing villages dotted here and there, blocking the way for bears eager to conquer new territories. This is a place where bears often wander across the schoolyard. Children are taught what to do if they see a bear, and they learn about the natural history of bears, without any fearmongering. This is a healthy example to follow.

Once in Slovenia, my colleagues and I were perched in an observation tower near a corn silo. We could hear horses whinnying in the distance and the sound of children playing in the schoolyard nearby. Suddenly, there was a shaking in the bushes, and a fine mother bear emerged, followed by one, two, three cubs. The family looked to be in good health, as the cubs were very playful. An anthill that had the misfortune to be standing close fell victim to their curiosity. Slovenia is home to one of the largest brown bear populations in Europe, and it's also a country of hunting, forestry, and livestock farming. There have always been bears here, and people have always adapted to their presence. Hunters pick off a few bears here and there; bears sometimes wreck a few beehives. When food is scarce in the mountains, they come down to the orchards to feast on apples and plums—that's how cohabitation works.

I've been traveling to Finland to observe bears for the last twenty years or so. Under the influence of the border authorities, observation towers with bear bait have been set up along the Russian border. Towers like these are not uncontroversial. Yes, they create an artificial environment, but without them it would be very difficult to observe bears exhibiting natural behavior such as mating, nursing their young, and marking their territory. These towers have enabled thousands of photographers, naturalists, and amateur observers to get up close with not only bears but also wolves and wolverines. The population is thriving here. Researchers have been observing bears for three decades, and records have been kept of the cubs that were born. Around a dozen of these observation centers are dotted throughout the historical area of Karelia, which straddles the modern-day border between Finland and Russia. Bear hunting is permitted outside these zones, but baiting bears with food has now been outlawed.

Similarly, in the United States, the Trump administration decided in 2017 to allow its hunting and livestock-farming friends to hunt bears again in and around Yellowstone Park, citing overpopulation—though fortunately, as mentioned earlier, that decision was overturned the next year. Across the border in Canada, grizzly bear hunting was made illegal in British Columbia in 2018. Political leaders have always told their electorate what they want to hear—it's a business relationship, a question of supply and demand. The power lies with us, the consumers, the citizens of the world, to dictate the decisions that are made. Bears and big carnivores should be indicators—and warning signs—of the health of our biodiversity.

These days, the status of bears has risen to a whole new level as individual bears are tagged and identified. Scientists and authorities are using more and more sophisticated technology in their quest to track, count, and study bears. One by one, bears in the wild are developing an identity of their own. In one case in the Pyrenees, a female and her two cubs were recorded as they emerged from their den after the winter by an automatic camera that now captures time-stamped video footage in increasingly high definition every time they pass by. Their scats were sampled and their genetic heritage tested to reveal which bear was the father of the cubs.

These cubs now have names, an estimated date of birth, a father, and photo ID. They were fitted with a collar so that researchers could track their every move by GPS for an entire year and study the data. The places they went, the naps they took, and the location of their winter den were all mapped out and confirmed by sampling in the field. If one of the bears is spotted or geolocated near livestock, or caught red-handed, it will get a black mark on its record and be flagged as a potential danger or troublemaker.

What will the future look like for large carnivores like brown bears? Human development continues to threaten their habitat, bear sanctuaries are not as idyllic a refuge as they were once thought to be, and their food sources are constantly being pillaged. The rate of deforestation in Eastern Europe is accelerating, for China's benefit. For now, the brown bear population across the world appears to be stable, but for how long will this be the case? Some say that bears have a place in zoos. Plenty of brown bear and polar bear cubs are being born in zoos and wildlife parks—a fancy term for the same kind of place—and newborn bear cubs are a goldmine for zoos. Animal welfare is also used as a justification. But if bears can't mate and reproduce normally, they must be chemically castrated to inhibit their hormonal cycle; otherwise they will develop stereotypic behavior, such as pacing and head swaying, to cope with their frustration. And if bears are left to live their regular lives, the cubs will keep on coming. No one will wonder about their future then—there will be too many and they'll have to be euthanized.

By way of conclusion

AS THE HUMAN population on our planet continues to increase and people grow hungrier for more land, forests, and hydroelectric dams, we need to rethink how all large carnivores, not just bears, fit into the big picture. Humans are going to come into contact—and conflict—with bears, wolves, lynxes, tigers, and other big cats more and more often. There are countless examples. Prides of lions have been poisoned by livestock farmers in Ethiopia, and cheetahs are dying out as demand for nature tourism grows. Wolves are being culled by the dozen in France, despite there being no serious grounds for regulating the population, and lynxes in the Vosges mountain range have been poisoned. And the list goes on. Large carnivores tend to clash with and magnify the resentment of people who, like them, live on the fringe of society. Unless concrete action is taken against the governments and elites exploiting them, these people will take their frustration out on the creatures that make them feel threatened. They don't realize they're doing all the dirty work for their masters. They are depopulating the wilderness of all the wild animals, making it easier for the authorities to acquire and exploit these areas in order to pave a highway or build a ski resort.

Peasants who kill bears in the Pyrenees or lions in Ethiopia are digging their own graves. It's only a matter of time before their days are numbered too. With the blessing of the Church, the great lords of the sixteenth and seventeenth centuries in Europe armed peasants so they could rid the land of large predators in order to make it easier to farm. Two centuries later, there were no more wild animals, and no more peasants either. If the people who live in areas that are still populated by bears, wolves, and tigers can truly value the biodiversity of their territory, they will keep the upper hand and be able to pave the way toward a new alliance. There is plenty to learn from reopening the history books and revisiting the battlegrounds.

History must be a weapon of mass construction. And for once, geography must be used not to wage war, but to repair the rift, so that city dwellers and country folk can look each other in the eye and rest easy knowing they understand one another. When all is said and done, doesn't coexisting with bears really mean learning to accept one another? It's simply a question of sharing what we have with another kind of being.

People who live for bears

Throughout my travels in search of bears, I've crossed paths with some interesting characters—people just as passionate about bears as I am, who have guided me along the way. I've encountered many scientists, of course, but also a fair few national park rangers, storytellers, and dreamers. What drives us all—or drove, in the case of those who are no longer with us—is our shared love of these great creatures.

For years, I had the pleasure of working alongside the American bear biologist Dr. Charles "Chuck" Jonkel (1930–2016) at the Churchill Northern Studies Centre in Manitoba. In his gravelly voice, he always used to talk about the first polar bear dens he discovered, in the 1960s. As the cofounder and president emeritus of the

Great Bear Foundation, Chuck was one of the pioneers of conservation in North America. A tireless public speaker, he also served as a mentor to many contemporary scientists.

At the Churchill Northern Studies Centre, I also came across Charlie Craighead, the son of Frank Craighead (1916–2001)—one of the famous Craighead twins, whose family has always been mad about bears and highly committed to grizzly conservation efforts in Yellowstone Park.

What characterizes these great North American naturalists is their enthusiasm to share their passion in all humility. These are just some of the other people who have made an impression on me through their discourse and their commitment:

In Canada: Ian Stirling, who pioneered the use of the polar bear tracking collar with Malcolm Ramsay (1949–2000), who lost his life in a helicopter accident.

In Russia: Igor Shpilenok, a photographer who lives for the Kamchatka Peninsula and its bears. He speaks to them, and they seem to listen. And there was Vitaly Nikolayenko, of course, as I mentioned earlier in this chapter.

In France: Jean-Jacques Camarra, who works as a researcher with the French National Hunting and Wildlife Agency (ONCFS) studying predators and pests, and Alain Reynes, director of the Association Pays de l'Ours-Adet, are both actively involved in the study and protection of bears in the Pyrenees. Each in their own way, they have contributed to a greater objective understanding of the population and the issues specific to these mountains.

And finally, all my good friends, guides, and partners in crime: Kevin Burke, Dennis Compayre, Mikhail Skopets, Kari Kemppainen, Jani Mataa, Konstantin, J.P., Marc, and others too numerous to mention.

We are all half-human and half-bear!

Laura Grizzlypaws, still image captured from the documentary *Fort comme un ours* (Strong like a bear), directed by Thierry Robert and Rémy Marion

Acknowledgments

FIRST OF ALL, I would like to thank François Sarano for setting the wheels of this book in motion by instigating my meeting with Stéphane Durand.

I am indebted to Stéphane for having worked my writing into his collection and for his invaluable comments.

All my gratitude goes to Françoise Passelaigue for her keen eye and attention to detail; thank you.

Of course I would like to thank my two partners in this life journey: my wife, Catherine, and son, Guillaume, alongside whom I work every day. They keep me on the straight and narrow, and without them I could never have done this job for the last twenty-five years.

Thank you to my colleague Farid Benhammou, brother of the bears.

Many thanks to the Cinquième Rêve and Arte team for their work on the production of my documentary *Fort comme un ours* (Strong like a bear). I must take my hat off to Thierry Robert, Aurélie Saillard, and Nicolas Zunino for all their encouragement.

All my thanks to Jeanne, Françoise, Patricia, and Alain at the Pôles Actions association, as well as all of our loyal supporters.

And I mustn't forget my travel companions: François, Delphine, Francis, Sylvie, Éliane, Raymond, Christian, Catherine, Georges, Francine, Olivier, Kristel, and all the others too numerous to mention here.

Notes

Introduction

1. John Muir, *A Thousand-Mile Walk to the Gulf* (Boston: Mariner Books, 1998; Boston: Houghton Mifflin, 1916).
2. Jean-Baptiste Charcot, *La Mer du Groenland* (Paris: Desclée de Brouwer, 1929). Excerpt translated from the original French.

Chapter 1: How to Describe a Bear

1. William Faulkner, *The Bear,* excerpted in *William Faulkner: The Critical Heritage,* ed. J. Bassett (London: Routledge, 2013). Used with the permission of Routledge, UK.
2. Georges-Louis Leclerc, Comte de Buffon, *Buffon's Natural History: Containing a Theory of the Earth, a General History of Man, of the Brute Creation, and of Vegetables, Minerals, &c. &c.,* vol. 6 (London: Symonds, 1797).
3. From the author's personal correspondence.
4. Jean Emmanuel Gilibert, *Abrégé du système de la nature de Linné* (Lyon: Matheron et Cie, 1802). Excerpt translated from the original French.

5. Jean-Jacques Camarra, *Boulevard des ours* (Toulouse, France: Milan, 1996). Excerpt translated from the original French.
6. Rick Bass, *The Lost Grizzlies: A Search for Survivors in the Wilderness of Colorado* (Boston: Houghton Mifflin, 1997).
7. François Merlet, *L'Ours, seigneur des Pyrénées* (Érables, 1988). Excerpt translated from the original French.
8. Doug Peacock, *Grizzly Years: In Search of the American Wilderness* (New York: Henry Holt, 1990).
9. Robert Hainard, *Mammifères sauvages d'Europe* (Paris: Delachaux et Niestlé, 1948). Excerpt translated from the original French.
10. Michel Pastoureau, *The Bear: History of a Fallen King*, trans. George Holoch (Cambridge, MA: Belknap Press of Harvard University Press, 2011).
11. You can view a slideshow of works by Éric Alibert, Robert Hainard, Philippe Legendre-Kvater, and Michel Bassompierre at vimeo.com/272757294.

Chapter 2: How to Become a Bear

1. Victor Hugo, *The Rhine*, trans. David Mitchell Aird (London, 1843).
2. A. C. Kitchener, "Taxonomic Issues in Bears: Impacts on Conservation in Zoos and the Wild, and Gaps in Current Knowledge," *International Zoo Yearbook* 44, no. 1 (January 2010): 33–46.
3. Video of polar bears feeding in Kaktovik, Alaska: vimeo.com/266321792.
4. Quoted in Sophie Bobbé, *L'Ours et le loup. Essai d'anthropologie symbolique* (Paris: Éditions de la Maison des sciences de

l'homme, 2002), 20. Excerpts translated from the original French.

5. Quoted in Bobbé, *L'Ours et le loup,* 20.
6. Video of polar bear cubs taking their first steps outside of the den, filmed in early March near Churchill, Manitoba: vimeo.com/65489167.

Chapter 3: How to Live Like a Bear

1. Baptiste Morizot, "Un seul ours debout," *Billebaude* no. 9 (September 2016). Excerpt translated from the original French.
2. Andy Russell, *Great Bear Adventures: True Tales From the Wild* (Toronto: Key Porter Books, 1994).
3. Valentin Pajetnov, *Avec les ours* (Arles, France: Actes Sud, 1998). Excerpt translated from the original French.
4. T. S. Smith and S. T. Partridge, "Dynamics of Intertidal Foraging by Coastal Brown Bears in Southwestern Alaska," *Journal of Wildlife Management* 68, no. 4 (2004): 233–40.
5. R. S. Waples, G. R. Pess, and T. Beechie, "Evolutionary History of Pacific Salmon in Dynamic Environments," *Evolutionary Application* 1, no. 2 (May 2008): 189–206.
6. Video of a brown bear fishing on the Kamchatka Peninsula: vimeo.com/190830510.
7. Video of brown bears in Kuhmo, Finland, near the Russian border, filmed in June: vimeo.com/69247827.
8. Guillaume Issartel, *La geste de l'ours: L'épopée romane dans son contexte mythologique XIIe-XIVe siècle* (Paris: Champion, 2010).

Chapter 4: The Bear in Its Environment

1. Camarra, *Boulevard des ours.*
2. Aldo Leopold, *A Sand County Almanac: And Sketches Here and There* (Oxford: Oxford University Press, 1968; first published 1949). All quotations taken from this work with the permission of Oxford University Press, USA.
3. G. V. Hildebrand, T. A. Hanley, C. T. Robbins, and C. C. Schwartz, "Role of Brown Bears (*Ursus arctos*) in the Flow of Marine Nitrogen Into a Terrestrial Ecosystem," *Oecologia* 121 (1999): 546–50.
4. Vladimir Arsenyev, *Across the Ussuri Kray: Travels in the Sikhote-Alin Mountains,* trans. J. C. Slaght (Bloomington, IN: Indiana University Press, 2016).
5. Peacock, *Grizzly Years.*
6. A. Tallian, A. Ordiz, M. C. Metz, C. Milleret, C. Wikenros, D. W. Smith, D. R. Stahler, J. Kindberg, D. R. MacNulty, P. Wabbaken, J. E. Swenson, and H. Sand, "Competition Between Apex Predators? Brown Bears Decrease Wolf Kill Rate on Two Continents," *Proceedings of the Royal Society B* 284, no. 1848 (2017).
7. Félix Maynard, *Un drame dans les mers boréales, souvenirs du Kamtschatka* (Paris: Michel Lévy frères, 1859). Excerpt translated from the original French.
8. Stepan Krasheninnikov, *The History of Kamtschatka, and the Kurilski Islands, With the Countries Adjacent,* trans. James Grieve (Gloucester, England: R. Raikes, 1764).

Chapter 5: The Bear's Winter of Mystery

1. Grey Owl, *Tales of an Empty Cabin* (Normanby Press, 2016).
2. G. V. Hildebrand, L. L. Lewis, J. Larrive, and S. D. Farley, "A Denning Brown Bear, *Ursus arctos,* Sow, and Two Cubs Killed in an Avalanche on the Kenai Peninsula, Alaska," *Canadian Field-Naturalist* 114, no. 3 (2000): 498.
3. A. Evans, N. J. Singh, A. Friebe, J. M. Arnemo, T. G. Laske, O. Fröbert, J. E. Swenson, and S. Blanc, "Drivers of Hibernation in the Brown Bear," *Frontiers in Zoology* 13, no. 7 (2016); M. Berg von Linde, L. Arevström, O. Fröbert, "Insights From the Den: How Hibernating Bears May Help Us Understand and Treat Human Disease," *Clinical and Translational Science* 8, no. 5 (2015): 601–5.
4. Nastassja Martin, "L'Ours," *Billebaude* 9 (September 2016). Excerpt translated from the original French.
5. Cornelius Osgood, *Ingalik Mental Culture* (New Haven: Yale University Publications in Anthropology, 1959), 146.
6. Johan Turi, *Turi's Book of Lappland,* trans. E. Gee Nash (New York: Harper & Brothers, 1931).
7. Leopold, *A Sand County Almanac.*

Chapter 6: The Geopoetic Polar Bear

1. Pierre Perrault, *Le Mal du Nord* (Gatineau, QC: Éditions Vents d'Ouest, 1999). Excerpt translated from the original French.
2. Image taken from Rémy Marion, *Sur les traces de l'ours polaire* (Tracking the polar bear), Nanouk Communication, 1999.

3. N. A. Øritsland, D. M. Lavigne, "Ultraviolet Photography: A New Application for Remote Sensing of Mammals," *Canadian Journal of Zoology* 52, no. 7 (1974): 939–41.
4. M. Q. Khattab and H. Tributsch, "Fibre-Optical Light Scattering Technology in Polar Bear Hair: A Re-Evaluation and New Results," *Journal of Advanced Biotechnology and Bioengineering* 13, no. 2 (2015).
5. Charcot, *La Mer du Groenland.*
6. Kenneth White, *Atlantica* (Paris: Grasset, 1986). Excerpt translated from the original French.

Chapter 7: The Bear's Revenge

1. Hainard, *Mammifères sauvages d'Europe.*
2. Stéphan Carbonnaux, *Le Cantique de l'ours. Petit plaidoyer pour le frère sauvage de l'homme* (Paris: Transboréal, 2008). Excerpt translated from the original French.
3. André Leroi-Gourhan, *Les Religions de la préhistoire*, Quadrige collection (Paris: Presses Universitaires de France, 1964). Excerpt translated from the original French.
4. Krasheninnikov, *The History of Kamtschatka.*
5. Jean-François Regnard, *Voyage en Laponie* (Paris, 1681). Excerpt translated from the original French.
6. Frederica de Laguna, Norman Reynolds, and Dale DeArmond, *Tales From the Dena: Indian Stories From the Tanana, Koyukuk & Yukon Rivers* (Seattle: University of Washington Press, 1995).
7. Marcel Couturier, *L'Ours brun* (Grenoble, 1954).
8. S. Erlinger, "Indications actuelles de l'acide ursodesoxycholique," *Hépato-Gastro & Oncologie Digestive* 1/9, no. 10

(December 2012): 817–22. Excerpt translated from the original French.

9. Berg von Linde et al., "Insights From the Den."
10. Peter Stenvinkel, "Studying Brown Bears to Treat Humans," Cité des sciences center in Paris, February 2–3, 2018, vimeo.com/255180016. Video in English, with French translation.
11. F. Sommer, M. Ståhlman, O. Ilkayera, J. M. Arnemo, J. Kindberg, J. Josefsson, C. B. Newgard, O. Fröbert, and F. Bäckhed, "The Gut Microbiota Modulates Energy Metabolism in the Hibernating Brown Bear *Ursus arctos*," *Cell Reports* 14, no. 7 (February 2016): 1655–61.
12. V. Tremaroli and F. Bäckhed, "Functional Interactions Between the Gut Microbiota and Host Metabolism," *Nature* 489 (September 2012): 242–49.

Chapter 8: Bears and Humans—the Future

1. Leopold, *A Sand County Almanac.*
2. Éric Baratay, *Animaux domestiques et animaux sauvages dans le discours clérical français des XVIIe-XVIIIe siècles,* in *L'homme, l'animal domestique et l'environnement du Moyen Âge au XVIIIe siècle,* ed. Robert Durand (Nantes, France: Ouest Éditions, 1993), 85–93.
3. Marquis Gaspard de Cherville, *Les Quadrupèdes de la chasse* (Nimes, France: Lacour, 2002). Excerpt translated from the original French.
4. Jim Harrison, *Just Before Dark: Collected Nonfiction* (New York: Open Road + Grove/Atlantic, 1993).
5. Romain Gary, *The Roots of Heaven,* trans. Jonathan Griffin (New York: Simon & Schuster, 1958).

6. Muir, A Thousand-Mile Walk to the Gulf.
7. Krasheninnikov, *The History of Kamtschatka*.
8. Nastassja Martin, "Vivre plus loin: Une rencontre d'ours chez les Even du Kamtchatka," *Terrain* 66 (October 2016): 142–55, doi.org/10.4000/terrain.16008. Excerpt used with the author's permission; translated from the original French. See also Nastassja Martin's book, *Croire aux fauves* (Paris: Éditions Gallimard, 2019).
9. F. Benhammou, S. Bobbé, J.-J. Camarra, and Alain Reynes, *L'Ours des Pyrénées, les 4 vérités* (Toulouse, France: Privat, 2005). Excerpt translated from the original French.
10. Sources: Ferus, ferus.fr.

Bibliography

Anon. *Tombent, tombent les gouttes d'argent. Chants du peuple aïnou*. L'aube des peuples collection. Paris: Gallimard, 1996.

Bernadac, Christian. *Le Premier Dieu*. Paris: Michel Lafon, 2000.

Brown, Gary. *Great Bear Almanac*. Guilford, CT: Lyons Press, 1993.

Camarra, Jean-Jacques. *L'Ours brun*. Paris: Hatier, 1989.

Caussimont, Gérard. *L'Ours brun des Pyrénées*. Carbonne, France: FIEP/Loubatières, 1997.

Crégut-Bonnoure, Evelyne. "The Saalian *Ursus thibetanus* From France and Italy." *Geobios* 30, no. 2 (1997): 285–94.

Dahl, Bjørn, and Jon E. Swenson. "Factors Influencing Length of Maternal Care in Brown Bears (*Ursus arctos*) and Its Effect on Offspring." *Behavorial Ecology and Sociobiology* 54, no. 4 (September 2003): 352–58.

Fagen, Robert, and Johanna M. Fagen, "Individual Distinctiveness in Brown Bears, *Ursus arctos*." *Ethology* 102, no. 2 (1996): 212–26.

Gastou, François-Régis. *Sur les traces des montreurs d'ours des Pyrénées et d'ailleurs*. Carbonne, France: Loubatières, 1987.

Goode, Erica. "Learning From Healthy Bears (You Mean We Should Hibernate?)." *New York Times*, July 4, 2016.

Grey Owl. *Tales of an Empty Cabin*. Normanby Press, 2016.

Grosse, C., P. Kaczensky, and F. Knauer. "Ants: A Food Source Sought by Slovenian Brown Bears (*Ursus arctos*)?" *Canadian Journal of Zoology* 81, no. 12 (2003): 1996–2005.

Hainard, Robert. *Choeur de loups et autres histoires d'ours*. Geneva, Switzerland: Slatkine, 1999.

Hoffman-Schickel, K., P. Le Roux, and É. Navet (eds.). *Sous la peau de l'ours: L'humanité et les Ursidés: approche interdisciplinaire*. Sources d'Asie collection. Saint-Denis, France: Connaissances et Savoirs, 2017.

Hoshino, Michio. *Grizzly*. San Francisco: Chronicle Books, 1987.

Kaczensky, Petra, Mateja Blazic, and Hartmut Gossow. "Public Attitude Towards Brown Bears (*Ursus arctos*) in Slovenia." *Biological Conservation* 118, no. 5 (August 2004): 661–74.

Kazeef, W. N. *L'Ours brun, roi de la forêt*. Paris: Stock, 1934.

Kryštufek, Boris, Bozidar Flajšman, and Huw I. Griffiths. *Living With Bears: A Large European Carnivore in a Shrinking World*. Ljubljana: Ecological Forum of the Liberal Democracy of Slovenia, 2003.

Kumar, V., F. Lammers, T. Bidon, M. Pfenninger, L. Kolter, M. A. Nilsson, and A. Janke. "The Evolutionary History of Bears Is Characterized by Gene Flow Across Species." *Scientific Reports* 7, art. 46487 (April 2017). doi: 10.1038/srep46487.

Kurtén, Bjørn. "Sex Dimorphism and Size Trends in the Cave Bear, *Ursus spelaeus* Rosenmüller and Heinroth." *Acta Zoologica Fennica* 90 (1955): 42–48.

Lajoux, Jean-Dominique. *L'Homme et l'ours*. Grenoble, France: Glénat, 1996.

Leroi-Gourhan, André. *Les Racines du monde. Entretiens avec Claude-Henri Rocquet*. Paris: Pierre Belfond, 1982.

Leroi-Gourhan, André, and Arlette Leroi-Gourhan. *Un voyage chez les Aïnous*, Paris: Albin Michel, 1989.

Lopez, Barry. *Arctic Dreams*. New York: Charles Scribner's Sons, 1986.

Lot-Falck, Éveline. *Les Rites de chasse chez les peuples sibériens*. Paris: Gallimard, 1953.

Loy, A ., P. Genov, M. Galfo, M. G. Jacobone, and A. Vigna Taglianti. "Cranial Morphometrics of the Apennine Brown Bear (*Ursus arctos marsicanus*) and Preliminary Notes on the Relationships With Other Southern European Populations." *Italian Journal of Zoology* 75, no. 1 (2008): 67–75.

Marion, Rémy. *Dernières nouvelles de l'ours polaire*. Barbizon, France: Pôles d'images, 2009.

Marion, Rémy, and Farid Benhammou. *Géopolitique de l'ours polaire*. Saint-Claude-de-Diray, France: Éditions Hesse, 2015.

de Marliave, Olivier. *Histoire de l'ours dans les Pyrénées*. Bordeaux, France: Éditions Sud-Ouest, 2000.

Matsuhashi, Tamako, Ryuichi Masuda, Tsutomu Mano, Koichi Murata, and Awirmed Aiurzaniin. "Phylogenetic Relationship Among Worldwide Populations of the Brown Bear *Ursus arctos*." *Zoological Science* 18, no. 8 (2001): 1137–43.

Mattson, David J., Colin M. Gillin, Scott A. Benson, and Richard R. Knight. "Bear Feeding Activity at Alpine Insect Aggregation Sites in the Yellowstone Ecosystem." *Canadian Journal of Zoology* 69, no. 9 (1991): 2430–35.

McCullough, D. R. "Population Dynamics of the Yellowstone Grizzly Bear." In *Dynamics of Large Mammal Populations*, edited by Charles W. Fowler and Tim D. Smith, 173–96. New York: John Wiley, 1981.

Miller, Susanne, James Wilder, and Ryan R. Wilson. "Polar Bear–Grizzly Bear Interactions During the Autumn Open-Water Period in Alaska." *Journal of Mammalogy* 96, no. 6 (November 2015): 1317–25.

Muir, John. *A Thousand-Mile Walk to the Gulf*. Boston: Mariner Books, 1998. First published in 1916 by Houghton Mifflin (Boston).

Nelson, Ralph A., G. Edgar Folk, Jr., Egbert W. Pfeiffer, John J. Craighead, Charles J. Jonkel, and Dianne L. Steiger. "Behavior, Biochemistry, and Hibernation in Black, Grizzly, and Polar Bears." *Bears: Their Biology and Management* 5 (1983): 284–90.

Noacco, Cristina. *Physiologos. Le bestiaire des bestiaires*. Translated, introduced, and with commentary by Arnaud Zucker. Grenoble, France: Éditions Jérôme Millon, 2004.

Parde, Jean-Michel, and Jean-Jacques Camarra. *L'ours (*Ursus arctos *Linnaeus, 1758)*. Vol. 5 of *Encyclopédie des carnivores de France:* Bourges, France: Société française pour l'étude et la protection des mammifères, 1992.

Passal, Jean-Noël. *L'Esprit de l'ours*. Le Coudray-Macouard, France: Cheminements, 2008.

Praneuf, Michel. *L'Ours et les hommes dans les traditions européennes*. Paris: Imago, 1989.

Prêtre, Bernard. *Les Derniers ours de Savoie et du Dauphiné*. Grenoble, France: Éditions de Belledonne, 2001.

Ramsay, Malcolm A., and Robert L. Dunbrack. "Physiological Constraints on Life History Phenomena: The Example of Small Bear Cubs at Birth." *The American Naturalist* 127, no. 6 (1986): 735–43.

Rockwell, David. *Giving Voice to Bear: North American Indian Myths, Rituals, and Images of the Bear*. New York: Roberts Rinehart Publishers, 1991.

Rozel, Ned. "Why Don't Hibernating Bears Get Osteoporosis?" UAF Geophysical Institute, art. 2234 (August 21, 2014). ktoo.org/2014/08/23/dont-hibernating-bears-get-osteoporosis/.

Schooler, Lynn. *L'Ours bleu*. Paris: Plon, 2002.

Shelton, James Gary. *Bear Attacks II: Myth and Reality*. Surrey, BC: Pallister Publications, 2001.

Sommer, F., M. Ståhlman, O. Ilkayeva, J. M. Arnemo, J. Kindberg, J. Josefsson, C. B. Newgard, O. Fröbert, and F. Bäckhed. "The Gut Microbiota Modulates Energy Metabolism in the Hibernating Brown Bear *Ursus arctos*." *Cell Reports* 14, no. 7 (February 2016). doi: 10.1016/j.celrep.2016.01.026.

Stenvinkel, P. J. Painer, M. Kuro-o, M. Lanaspa, W. Arnold, T. Ruf, P. G. Shiels, and R. J. Johnson. "Novel Treatment Strategies for Chronic Kidney Disease: Insights From the Animal Kingdom." *Nature Reviews Nephrology* 14, no. 4 (2018): 265–84.

Tait, David E. N. "Abandonment as a Reproductive Tactic: The Example of Grizzly Bears." *American Naturalist* 115, no. 6 (June 1980): 800–8.

Turi, Johan. *Récit de la vie des Lapons*. Paris: L'Harmattan, 1997.

Walter, Philippe. *Arthur, l'ours et le roi*. Paris: Imago, 2002.

Wang, Xiaoming, Natalia Rybczynski, C. Richard Harington, Stuart C. White, and Richard H. Tedford. "A Basal Ursine Bear (*Protarctos abstrusus*) From the Pliocene High Arctic Reveals Eurasian Affinities and a Diet Rich in Fermentable Sugars." *Scientific Reports* 7, art. 17722 (2017). doi: 10.1038/s41598-017-17657-8.

Welinder, K. G., R. Hansen, M. T. Overgaard, M. Brohus, M. Sønderkær, M. von Bergen, U. Rolle-Kampczyk, et al. "Biochemical Foundations of Health and Energy Conservation in Hibernating Free-Ranging Subadult Brown Bear *Ursus arctos*." *Journal of Biological Chemistry* 291, no. 43 (October 21, 2016): 22509–523.

Xenikoudakis, G., E. Ersmark, J.-L. Tison, L. P. Waits, J. Kindberg, J. E. Swenson, and L. Dalén. "Consequences of a Demographic Bottleneck on Genetic Structure and Variation in the Scandinavian Brown Bear." *Molecular Ecology* 24, no. 13 (July 2015): 3441–54.

Additional Resources

Selected books by Rémy Marion

Ovibos: Le survivant de l'arctique, with Robert Gessain (Arles, France: Actes Sud, 2020). Available in French.

Penguins: A Worldwide Guide (New York: Stirling, 1999).

On the Trail of Bears, with Catherine Marion (Hauppauge, New York: Barron's Educational Series, 1998).

On the Trail of Whales, with Jean-Michel Dumont (Hauppauge, New York: Barron's Educational Series, 1998).

Films by Rémy Marion available in English

The Superpowers of the Bear, co-directed with Thierry Robert (France: Arte / Le Cinquième Rêve, 2018).

Evolution of the Polar Bear, co-directed with Charlène Gravel (France: Arte / Bonne Pioche / Pôles d'images, 2015).

Visit Poles D'images on Vimeo at vimeo.com/user16797628 for talks and videos related to bears and other wildlife.

Visit pacificwild.org for information and wildlife protection campaigns in British Columbia's Great Bear Rainforest.

About the Pôles Actions association

The Pôles Actions association was founded to foster awareness and promote the protection of the Arctic and Antarctic ecosystems through information, education, and support for research in the polar and subpolar regions. The association is working to open an objective, non-partisan, and scientifically credible channel of communication that draws on the activities and experience of its members and the external stakeholders it calls upon. Pôles Actions organizes educational events, scientific talks, and exhibitions with a view to cultivating knowledge and protecting these fragile ecosystems. All forty-five talks presented at the four conferences the association has organized are available online (in French only) at ourspolaire.org.

Image Credits

page xii: © Rémy Marion

pages 29 and 131: © Éric Alibert

page 170: © Patrick Aventurier, Chauvet–Pont d'Arc

page 203: © Le Cinquième Rêve / Arte / Pôles d'images

Index

Note: Page numbers in italics refer to images and maps.

RÉMY MARION is an author, photographer, and documentary filmmaker who has devoted his life to observing bears in the wild since the 1980s. Marion is a member of France's national societies of geographers and explorers, and he is considered to be one of the country's leading authorities on brown bears and polar bears.

LAMBERT WILSON is an actor and an environmental and human rights activist best known internationally for his roles in the Matrix film franchise and in the 2016 film *The Odyssey*.